Business Continuity Plan (BCP) Te
With Instructions and Exampl

- Erik Kopp, MBA

CW01667604

This book provides information on how to set up, use, and maintain a Business Continuity Plan (BCP). Starting with the blank BCP Master Template, the instructions provide guidance on how to create a final workable BCP plan that you can use in the event of an emergency or unplanned interruption to your business. The master template is a starting point which is intended to be customized to your individual business needs.

The template is formatted in individual sections for each phase of the BCP process and for each critical business process. This may be used in its entirety, or it can be easily broken apart and distributed as separate documents to the appropriate individuals who need them.

The book includes 3 parts:

1. A blank BCP Master template which you can copy and fill in to use as needed.
2. Instructions for Using, and Maintaining A Business Continuity Plan (BCP).
3. An example of a completed BCP plan, to illustrate how it can be used.

To obtain an electronic copy of the BCP Master Template shown in this book completely free, please go to http://www.erikkopponline.com/bcptemplate.html (password "Sunday") to download the file. Please note that this requires WORD 2003 or later to use the template.

Published By:
EK Publications
http://smallbusinesscontinuity.weebly.com/

Table of Contents

PART 1 – BCP MASTER TEMPLATE.....5

SECTION I - PLAN OVERVIEW AND CONTACT INFORMATION.....5
1. Plan Summary.....5
2. Plan Approval.....5
3. BCP Leadership Roles and Responsibilities.....6
4. Internal Contact Information.....9
5. External Contact Information (Customers).....13
6. External Contact Information (Suppliers and Business Service Providers).....14
7. Utilities and Facilities Services.....16
8. Financial Services Contact Information.....18
9. Regulatory Agency Contact Information.....20
10. Critical Records and Systems.....21
11. Backup Locations.....22
12. Backup Service Provider and Supplier Information.....23
13. Emergency Services Contact Information.....24
SECTION II - BUSINESS RISK ASSESSMENT AND IMPACT EVALUATION.....25
14A. Business Process Risk Evaluation.....26
14B. Business Process Disruption Impact Evaluation.....27
SECTION III - PRE-EMERGENCY PLANNING.....28
15. Risk Mitigation.....28
16. Emergency Materials and Supplies.....29
SECTION IV - BCP PLAN ACTIVATION AND IMPLEMENTATION.....30
17. Plan Activation.....30
18. Plan Implementation.....31
SECTION V - BCP PLAN DEACTIVATION AND RESTORATION OF NORMAL OPERATIONS.....33
19. BCP Plan Deactivation.....33
20. Return To Normal Operations.....34
SECTION VI - BCP PLAN TESTING.....36
SECTION VII - BCP PLAN MAINTENANCE.....37
SECTION VIII - APPENDICES / ATTACHMENTS.....38
APPENDIX ___ - <INSERT TITLE HERE>.....38

PART 2 – INSTRUCTIONS FOR SETTING UP AND MAINTAINING THE BCP PLAN.....39

SECTION I - PLAN OVERVIEW AND CONTACT INFORMATION.....39
1. Plan Summary.....39
2. Plan Approval.....39
3. BCP Leadership Roles and Responsibilities.....39
4. Internal Contact Information.....41
5. External Contact Information (Customers).....43
6. External Contact Information (Suppliers and Business Service Providers).....43
7. Utilities and Facilities Services.....43
8. Financial Services Contact Information.....43
9. Regulatory Agency Contact Information.....44
10. Critical Records and Systems.....44
11. Backup Locations.....45
12. Backup Service Provider and Supplier Information.....45
13. Emergency Services Contact Information.....46
SECTION II - BUSINESS RISK ASSESSMENT AND IMPACT EVALUATION.....47
14A. Business Process Risk Evaluation.....47
14B. Business Process Disruption Impact Evaluation.....48
SECTION III - PRE-EMERGENCY PLANNING.....49
15. Risk Mitigation.....49
16. Emergency Materials and Supplies.....49
SECTION IV - BCP PLAN ACTIVATION AND IMPLEMENTATION.....50
17. Plan Activation.....50

18. Plan Implementation ... *50*
SECTION V - BCP PLAN DEACTIVATION AND RESTORATION OF NORMAL OPERATIONS ... 51
19. BCP Plan Deactivation ... *51*
20. Return To Normal Operations ... *51*
SECTION VI - BCP PLAN TESTING ... 52
SECTION VII - BCP PLAN MAINTENANCE ... 53
SECTION VIII - APPENDICES / ATTACHMENTS ... 54

PART 3 - EXAMPLE: BUSINESS CONTINUITY PLAN (BCP) FOR EK CATERING SERVICE, LLC ... 55

SECTION I - PLAN OVERVIEW AND CONTACT INFORMATION ... 56

1. PLAN SUMMARY ... 56
2. PLAN APPROVAL ... 56
3. BCP LEADERSHIP ROLES AND RESPONSIBILITIES ... 57
4. INTERNAL CONTACT INFORMATION ... 60
5. EXTERNAL CONTACT INFORMATION (CUSTOMERS) ... 65
6. EXTERNAL CONTACT INFORMATION (SUPPLIERS AND BUSINESS SERVICE PROVIDERS) ... 67
7. UTILITIES AND FACILITIES SERVICES ... 69
8. FINANCIAL SERVICES CONTACT INFORMATION ... 70
9. REGULATORY AGENCY CONTACT INFORMATION ... 71
10. CRITICAL RECORDS AND SYSTEMS ... 72
11. BACKUP LOCATIONS ... 74
12. BACKUP SERVICE PROVIDER AND SUPPLIER INFORMATION ... 75
13. EMERGENCY SERVICES CONTACT INFORMATION ... 76

SECTION II - BUSINESS RISK ASSESSMENT AND IMPACT EVALUATION ... 77

14A. BUSINESS PROCESS RISK EVALUATION ... 78
14B. BUSINESS PROCESS DISRUPTION IMPACT EVALUATION ... 79

SECTION III - PRE-EMERGENCY PLANNING ... 82

15. RISK MITIGATION ... 82
16. EMERGENCY MATERIALS AND SUPPLIES ... 83

SECTION IV - BCP PLAN ACTIVATION AND IMPLEMENTATION ... 84

17. PLAN ACTIVATION ... 84
18. PLAN IMPLEMENTATION ... 86

SECTION V - BCP PLAN DEACTIVATION AND RESTORATION OF NORMAL OPERATIONS ... 91

19. BCP PLAN DEACTIVATION ... 91
20. RETURN TO NORMAL OPERATIONS ... 94

SECTION VI - BCP PLAN TESTING ... 98

SECTION VII - BCP PLAN MAINTENANCE ... 99

SECTION VIII - APPENDICES / ATTACHMENTS ... 100

APPENDIX I - MAP AND DRIVING DIRECTIONS TO ACE KITCHENS INC. ... 100
APPENDIX II - INSTRUCTIONS FOR LOADING AND UNLOADING THE FREEZER AT GB BAKERY ... 101

New Material for 2nd Edition

- BCP Essentials (Key Requirements for Effective BCPs)

- Personal Emergency Plan for Yourself and Your Family

PART 1 – BCP MASTER TEMPLATE

Business Continuity Plan (BCP) for <Company Name>

Section I - Plan Overview and Contact Information

1. Plan Summary

This Business Continuity Plan (BCP) has been developed to ensure <Company Name> is prepared to respond to unplanned business interruptions that may affect critical business processes. An unplanned business interruption is defined as an event (such as an emergency or disaster) that causes an unanticipated disruption to the operation of the business and/or service to customers. The objective of this plan is to minimize the impact of unplanned interruptions and provide guidance for continuing operations during an unplanned event and resuming normal operations once the event is resolved.

2. Plan Approval

All BCP plans must be approved by at least one individual in a senior company management role.

	Printed Name	Title	Signature	Date
2.1 Prepared By:				
2.2 Approved By:				
2.3 Approved By:				

3. BCP Leadership Roles and Responsibilities

BCP leadership roles and responsibilities must be identified and defined before an emergency occurs in order for the BCP plan to be effective.

BCP ID	Role	BCP Responsibility	Name
3.1	Company Management / Site Management (Primary)	Activate BCP plan upon notification of emergency, disaster or unplanned event causing interruption to business operations. Deactivate the BCP plan when conditions allow for transition back to normal operations.	
3.2	Company Management / Site Management (Backup)	Backup person who can activate the BCP in the event that the primary person responsible for BCP activation is unavailable.	
3.3	BCP Team Leader (Primary)	Provide leadership and decision making authority to BCP team upon activation of the BCP plan, until plan is deactivated.	
3.4	BCP Team Leader (Backup)	Backup person to serve as BCP leader in the event that primary BCP leader is unavailable.	

BCP ID	Role	BCP Responsibility	Name
3.5	BCP Communications Leader (Primary)	Manage external communications between the company and outside parties and internal communications between the BCP team and company personnel.	
3.6	BCP Communications Leader (Backup)	Backup person to serve as Communications Leader in the event that primary Communications Leader is unavailable.	
3.7	Plan Administrator (Primary)	Own and mange the BCP plan to ensure it is current and accessible in the event of an emergency. Ensure review and testing is performed and documented on a regular basis.	
3.8	Plan Administrator (Backup)	Backup person to serve as Plan Administrator in the event that primary Plan Administrator is unavailable.	
3.9	Functional Lead (Primary) <Function1>	Functional leader/manager in charge of <a business operation or process (i.e. purchasing, manufacturing, distribution, IT, etc)>	

BCP ID	Role	BCP Responsibility	Name
3.10	Functional Lead (Backup) <Function1>	Backup person to serve as Functional leader/manager for <business operation>in the event that primary Functional leader/manager is unavailable.	
3.11	<please expand this list to include primary and backup leaders for all critical business functions identified>		

4. Internal Contact Information

In the event of an emergency or disaster, it is critical to have contact information for all personnel to be able to quickly share important information as it becomes available and to quickly account for all personnel. The most useful format for this information, in the event of an emergency, is to set it up as one or multiple contact trees (depending on how big your organization is).

The contact tree is a structured hierarchical format with multiple levels which look like a pyramid if you draw them in a diagram (see example below).

In the event of an emergency, the BCP leadership contacts the person at the top of the tree (Level 1). The Level 1 person then contacts the assigned people in the tree one level down (Level 2). The Level 2 people then contact the people assigned to them one level down (Level 3). Once the Level 2 people have contacted (or attempted to contact) all the Level 3 people assigned to them, they then contact the Level 1 person above them to report the status (All People Accounted For or People Not Accounted For) and any question or information to be shared. This provides confirmation that the contact tree has reached all the affected individuals and that everyone is accounted for. The Level 1 person at the top of the tree is then responsible for reporting back to the BCP Leadership the status of people in their contact tree and conveying any information they need to share.

The contact tree can be set up various ways depending on what makes sense to your organization such as supervisor with associated personnel below, or it make more sense to have a level 1 site person with assigned people at the same site below (for organizations where reporting structures span multiple locations).

The contact tree can be expanded (more levels added) for larger groups. Multiple contact trees can be set up for multiple groups, functions, or locations.

The table provided below is set up in a contact tree structure to identify each person within a contact tree and to provide instructions on what to do in the event the contact tree is activated. Information about each person should include how to reach them through multiple means (phone, email, text, etc) and at different times (work hours, after work hours, etc). It is also very useful to note any special skills or training each person has (i.e. paramedic, firefighter, doctor, nurse, etc.) in the event their services are needed in an emergency.

Example Contact Tree Diagram

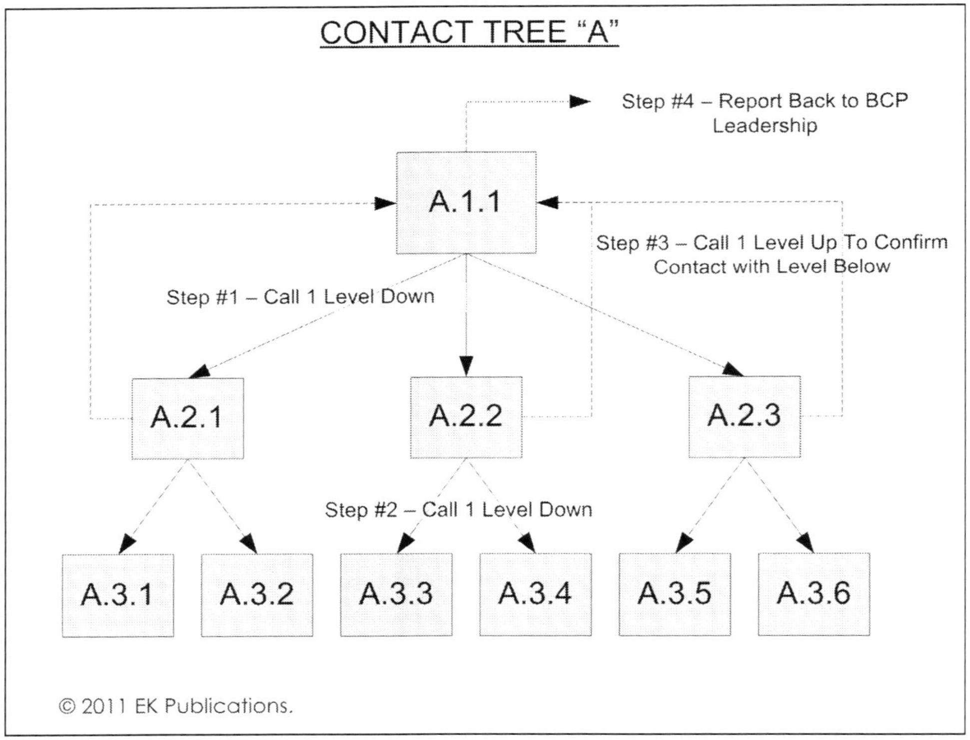

Contact tree ID	Name	Title and Affiliation (Employee, Contractor)	Emergency Skills/Qualifications (ex. EMT, MD, RN)	(W)ork Phone/ (H)ome Phone/ (C)ell Phone	email	Contact tree Instructions
<CONTACT TREE A – TITLE ex. Field Sales> - Level 1						
A.1.1	Stan Brown			(W) (H) (C)		Ex. - Contact A.2.1___ - Contact A.2.2___ - Contact A.2.3___ - Confirm contacts complete with BCP Leadership upon notification from A2.1, A2.2___
<CONTACT TREE A – TITLE ex. Field Sales> - Level 2						
A.2.1	Sally Jones					Ex. - Contact A.3.1___ - Contact A.3.2___ - Confirm contacts complete with A.1.1___
A.2.2	Phil Smith					- Contact A.1.1 as needed.
A.2.3	Jim White					- Contact A.3.3___ - Confirm contacts complete with A.1.1___
<CONTACT TREE A – TITLE ex. Field Sales> - Level 3						

Contact tree ID	Name	Title and Affiliation (Employee, Contractor)	Emergency Skills/Qualifications (ex. EMT, MD, RN)	(W)ork Phone/ (H)ome Phone/ (C)ell Phone	email	Contact tree Instructions
A.3.1	Jane Wilson					- Contact A.2.1 as needed.
A.3.2	Kate Green					- Contact A.2.1 as needed.
A.3.3	Linda Hudson					- Contact A.2.3 as needed.
A.3.4	<please expand this list to include contact information for all internal personnel affected by this BCP plan>					
<CONTACT TREE B – TITLE > - Level 1						
B.1.1	<please expand this list to include contact information for all internal personnel affected by this BCP plan>					

5. External Contact Information (Customers)

A critical piece of information to have available in the event of an emergency causing interruptions to business operations is a customer list to enable timely communications to customers during outages. This list can either be compiled using the table below, or if customer information is maintained in a separate system and a list can be printed or copied and pasted, attach it to this BCP plan as an Appendix and reference it here in place of the table.

BCP ID	Name	Title	Company Name and Address	(W)ork Phone/ (H)ome Phone/ (C)ell Phone	Web address, email
5.1				(W) (H) (C)	
5.2					
5.3					
5.4					
5.5					
5.6					
5.7					
5.8					
5.9	<please expand this list to include contact information for all customers affected by this BCP plan>				

6. External Contact Information (Suppliers and Business Service Providers)

A critical piece of information to have available in the event of an emergency causing interruptions to business operations is a list of suppliers and service providers to enable timely restoration of services or implementation of backup services during an interruptions to normal business operations. This list can either be compiled using the table below, or if supplier and service provider information is maintained in a separate system and a list can be printed or copied and pasted, attach it to this BCP plan as an Appendix and reference it here in place of the table.

BCP ID	Service Provided	Company	Name / Title	Account / Policy #	Phone	email
6.1	Insurance					
6.2	Legal Services					
6.3	Computer Services / Web Site Hosting					
6.4	Telephone Company (landline)					
6.5	Cell Phone Provider					
6.6	Internet Service Provider					
6.7	Cable / Satellite Company					

BCP ID	Service Provided	Company	Name / Title	Account / Policy #	Phone	email
6.8	Shipping Company					
6.9	Material Suppliers					
6.10	Travel Agency					
6.11	Site Maintenance Company					
6.12	<please expand this list to include contact information for all service providers and suppliers necessary to continue business operations>					

7. Utilities and Facilities Services

A critical piece of information to have available in the event of an emergency causing interruptions to business operations is a list of utility and site maintenance companies to enable timely restoration of services or implementation of backup services during an interruption to normal business operations, or to allow for timely notification to cut service to remove unsafe conditions at the work site (i.e. gas leak, downed power wires, etc.). This list can either be compiled using the table below, or if utility and site management information is maintained in a separate system and a list can be printed or copied and pasted, attach it to this BCP plan as an Appendix and reference it here in place of the table.

BCP ID	Service Provided	Company	Name	Title	Phone	email
7.1	Landlord / Leasing Company / Site Owner					
7.2	Gas Company					
7.3	Electric company					
7.4	Water Company					
	PO Box Rental Company					

BCP ID	Service Provided	Company	Name	Title	Phone	email
7.5	<please expand this list to include contact information for all utilities necessary to continue business operations>					

8. Financial Services Contact Information

In the event of an emergency or disruption to business requiring relocation, it is imperative to maintain contact with financial institutions to enable continuation of payments to employees, suppliers and service providers, as well as processing payments from customers. This list can either be compiled using the table below, or if financial institution information is maintained in a separate system and a list can be printed or copied and pasted, attach it to this BCP plan as an Appendix and reference it here in place of the table. This information should also identify key individuals who have knowledge of account numbers and have system IDs and passwords allowing access to financial data and services. It is not recommended to publish this information within the BCP plan as this may result in unauthorized access. The plan should specify a primary and secondary individual responsible for accessing financial data and services.

BCP ID	Financial Service	Contact Names (Primary and Secondary)	Address	Phone	email	Primary and Secondary Individuals with access to account, ID and password/PIN (specify Contact Tree ID from section I, 4)
8.1	Bank					<Ex. Jane Smith (Primary), BCP ID 3.2 Kate Wilson (Secondary), BCP ID 3.12>
8.2	Payroll Administrator					
8.3	Benefits Administrator					

BCP ID	Financial Service	Contact Names (Primary and Secondary)	Address	Phone	email	Primary and Secondary Individuals with access to account, ID and password/PIN (specify Contact Tree ID from section I, 4)
8.4	<please expand this list to include contact information for all appropriate financial services>					

9. Regulatory Agency Contact Information

It is critical to maintain communications with regulatory agencies during emergencies to enable reporting of business status affecting required regulatory filings or reporting.

BCP ID	Regulatory Agency	Contact Names (Primary and Secondary)	Address	Phone	email
9.1	IRS				
9.2	State Tax Dept.				
9.3	Social Security				
9.4	Licensing Bureau (if applicable)				
9.5	SEC (if applicable)				
9.6	FDA (if applicable)				
9.7	<please expand this list to include contact information for all appropriate regulatory agencies>				

10. Critical Records and Systems

To ensure continuation of the business with minimal interruptions or negative impact to your customers, it is important to have access to critical information such as customer lists, orders, financial statements, personnel records, tax records, contracts, etc. These may reside in hard copy form or in electronic systems. For all of these records, regardless of the media or format, there must be backup copies available in the event of en emergency. Back up copies may be paper copies, microfilm, scanned files, electronic copies (CDs, DVDs, Memory stick etc.) or these may be maintained in computerized systems with remote access (i.e. Online Mailbox, online servers, etc.). The critical piece is that in the event of an emergency you can access the information you need quickly and from an alternate location. For this section, document each critical record (invoices, POs, contracts, etc.) and system (market research database, order processing system, web site server, etc.) and identify its backup. If you were to lose primary access to this information, where would you go to find it? This list should contain information needed to get to the back up record location or system (what do I need to know to be able to find this information in an emergency). It is not necessary to document system IDs and passwords here (unless you want to) but at a minimum you need to identify who has the necessary system IDs and passwords, pin, keys, combinations, etc. to be able to access the records/systems. Be sure to identify a primary and backup person here.

BCP ID	Critical Record or System	Backup Record Location / System (URL, Name&Address – how to find this information)	Primary Person With Access To This Information (Name, Contract Tree ID)	BackUp Person With Access To This Information (Name, Contract Tree ID)
10.1				
10.2				
10.3				
10.4				
10.5	<please expand this list to include contact information for all Critical Records and Systems>			

11. Backup Locations

To ensure continuation of critical business functions during activation of the BCP plan, backup/contingency locations must be identified for all essential operations which cannot be performed remotely (at home). Examples are laboratory or manufacturing operations, and conference facilities for meeting with customers. Backup sites should be far enough away from the primary work site to reduce the chances that the disaster will affect access to the backup site, but not so far as to prevent timely access by personnel. A general rule is for the backup site to be 20-50 miles from the primary site. Backup site information in the BCP plan must include directions instructing personnel how to get to the location. This list can either be compiled using the table below, or if backup location information is maintained in a separate system and a list can be printed or copied and pasted, attach it to this BCP plan as an Appendix and reference it here in place of the table.

BCP ID	Function / Operation	Backup Location Name*	Contact Names (Primary and Secondary)	Address	Phone	email
11.1						
11.2						
11.3						
11.4						
11.5	<please expand this list to include contact information for all appropriate backup locations>					

* **Directions** and **Maps** to all backup locations are included in the Appendix Section of this BCP.

12. Backup Service Provider and Supplier Information

In the event of an emergency or disruption to primary business services and suppliers, a list of back up services and suppliers must be maintained for quick reference to enable restoration of essential business functions in a timely manner. This list can either be compiled using the table below, or if backup services and supplier information is maintained in a separate system and a list can be printed or copied and pasted, attach it to this BCP plan as an Appendix and reference it here in place of the table.

BCP ID	Service / Supplier	Company	Contact Names (Primary and Secondary)	Address	Phone	email
12.1	<ex. Raw Materials, Manufacturing, Printing, Shipping, Order processing>					
12.2	Computer services / web hosting					
12.3	Payroll administration					
12.4	File archives					
12.5	<please expand this list to include contact information for all appropriate backup service providers and suppliers>					

13. Emergency Services Contact Information

In any emergency situation, it is essential to know who to contact. This information must be on hand for easy availability. List all critical emergency contact information in the table below so you have it all in once place for easy access.

BCP ID	Emergency Service	Contact Name(s) (if it is important)	Address (if it is important)	Phone (include direct numbers if known, in case 911 services are unavailable)	Email / Web
13.1	Local Police Dept.				
13.2	State Police				
13.3	Fire Dept.				
13.4	Local Emergency Management Director				
13.5	First Aid Squad				
13.6	Hazmat				
13.7	FBI				
13.8	Homeland Security				
13.9	Poison Control				
13.10	Local Township/Municipality				
13.11	<please expand this list to include contact information for all appropriate emergency services>				

Section II - Business Risk Assessment and Impact Evaluation

This section of the BCP plan documents the risk associated with the disruption of any key business operation or process. Risk is defined as the combination of how likely a key operation will be disrupted, how much time before the business experiences the negative impact of losing the operation, and how much this disruption will hurt the business performance. (*Ex. The manufacturing plant can be shutdown for several months in the event that the river running through the property overflows and floods the building. Some flooding has occurred once every 3 years with major flooding occurring on average every 25 years. This will result in loss of sales within 2 days*).

The information documented in this section is used to justify the prioritization of the BCP plan activities and justify the money allocated to ensuring the continued operations during an emergency or disaster. (*Ex. A backup manufacturing site is qualified and PO-##### is issued to Company Z to enable start up of production within 1 day in the event of an emergency at the primary manufacturing facility. A new manufacturing site has been identified and a project initiated to relocate primary manufacturing to this site by 15-JUN-20##*).

The more these figures can be quantified, the better will be the BCP decision-making.
To assist in this evaluation, use the scale indicated in the chart in Section 14.

Once this section is complete, this information will provide a roadmap for prioritizing the BCP tasks.
The result will be a list of key business operations/processes with risk levels assigned. The operations/processes with the lowest numbers will have the highest risk level. The high risk items are the ones which should receive the most attention when designing and implementing the BCP plan. These are the items which will hurt the business the most if they are not adequately addressed.

14A. Business Process Risk Evaluation

The critical elements of completing this section are:

- Identifying the key business operations and processes. These are operations or process which must be working in order for the business to continue.
- Identifying the likelihood of disruption (chances that something unexpected could happen which disrupts the key business operation or process). **SCALE: 1 – 10** (1 = Very High Likelihood of Unplanned Disruption, 10 = Very Low Likelihood of Unplanned Disruption)
- Identifying how significant the effect is of losing this operation or process. (i.e. people at risk, profits lost, customers lost) **SCALE: 1 – 10** (1 = Very High Significance To The Business, 10 = Very Low Significance To The Business).
- Identifying how long this operation can be disrupted before it negatively impacts the business. **Number of Days** without this operation before business would experience negative impact to operations and/or sales.

BCP ID	**Key Business Process (Operation/Process which must be working for the business to continue)**	**(A) Likelihood Of Disruption SCALE: 1 (High) - 10 (Low)**	**(B) Significance Of Negative effect if disrupted for more days than identified in (C) SCALE: 1 (High) - 10 (Low)**	**(C) Time Before Negative Business Effect (Number of Days)**	**BCP Priority = (A) x (B) + (C) (Lower Number = Higher Risk and Higher BCP Priority)**
14.1					
14.2					
14.3					
14.4					
14.5	<please expand this list to include information for all key business processes>				

14B. Business Process Disruption Impact Evaluation

For the key business processes identified in section 14A, evaluate the impact to the business of an unplanned disruption in terms of:

- Available work-arounds to temporarily replace the process *(ex. paper records can be processed manually in the event that a key computer system is unavailable, process can be outsourced to Company X).*
- How long can the business continue to operate until the process must be restored? *(ex. computer system must be back online within 3 months or costs and loss of efficiency will reach unsustainable levels).*
- What resources are needed for work-around and for restoration of the process? *(ex. for the duration of manual processing of paper forms, resources needed include: 3 temporary people working 8 hours per day each, 3 computers with internet access, 1 printer, XYZ software with 3 licenses, workspace for 3 people, 1 telephone, 1 FAX, etc.).*

List the Key Business Processes in order of BCP Priority (lowest number first) using the information identified in the table in question 14A. The objective here is to identify which processes require the most immediate attention in the event of a disruption and what is required to address the disruption.

BCP ID (From section 14A.)	Key Business Process	Work-arounds available to temporarily replace this process.	How long can business operate until process must be restored? (Number of Days)	Resources needed to temporarily replace this process. (People, Services, Equipment, Facilities, Money, etc.)	Resources needed to restore this process. (People, Services, Equipment, Facilities, money, etc.)
14.1					
14.2					
14.3					
14.4					
14.5	<please expand this list to include information for all key business processes>				

Section III - Pre-Emergency Planning

15. Risk Mitigation

Identify any risk mitigation options are available to reduce the chances of an unplanned disruption occurring and/or reduce the negative impact to the business. *(Ex. Create 2 backup copies of sales forecast and customer database and store in 2 different secure offsite locations. Location 1 is __________, location 2 is _________).*

BCP ID	**Key Business Process**	**Risk Mitigation Plan (if none possible, indicate "none")**
15.1		
15.2		
15.3		
15.4		
15.5	<please expand this list to include information for all key business processes>	

16. Emergency Materials and Supplies

For each Key Business Process, identify all critical emergency materials and supplies that need to be available in the event of an emergency or unplanned business interruption. This includes small disposable supplies (i.e. pens, paper, mailing supplies, etc.), office equipment (computers, printers, phones, etc.), production equipment (i.e. machinery needed to service customers, refrigerator, freezer, generator, batteries, tools, etc.) as well as raw materials and production supplies (i.e. paper, paint, lawn fertilizer, baking ingredients, gasoline, propane, etc.).

BCP ID	Key Business Process	Materials / Supplies needed (Location or Contact information)
16.1		
16.2		
16.3		
16.4		
16.5	<please expand this list to include information for all key business processes>	

Section IV - BCP Plan Activation and Implementation

In this section, identify the triggers for activating the BCP plan (what events must occur in order to activate the BCP plan), who is involved in activating the BCP plan, and what are the steps to activate and implement the plan. All key business processes identified in Section 14A should be listed in Section 16 in order of BCP priority.

Start times should be specified based on impact evaluation information captured in section 14B (*ex. if production line downtime causes loss of customer orders within 2 days, then the trigger is that production line is down for more than 1 day and the target start time for production operation BCP should be specified to have a backup process in place within 2 days*).

17. Plan Activation

BCP ID	Key Business Process	BCP Activation Trigger	Responsible Individual(s) for Activating / Implementing BCP	BCP Plan ID (refer to section 18)	BCP Contact Tree* (refer to sections 3 and 4)
17.1					
17.2					
17.3					
17.4					
17.5	<please expand this list to include information for all key business processes>				

*Upon activation of the BCP plan initiate the contact tree (Refer to Sections 3 and 4) to account for all personnel and to alert team members of BCP activation

18. Plan Implementation

In this section, define the tasks needed to start operating under the BCP plan. For each task, document who is assigned the responsibility for completing the task. The tables below are set up to document the tasks, who is responsible and a section for verification that the task was completed. As an alternative to these tables, the implementation plan can be a separate document attached to this document as an Appendix.

18.2 Key Business Process: ____________________

Start Time from BCP Activation (Days, Hours)	BCP Task	Assigned To	Completion Verification (enter Initials / Date / Time upon completion)
			______ / ______ / ______

18.2 Key Business Process: ____________________

Start Time from BCP Activation (Days, Hours)	BCP Task	Assigned To	Completion Verification (enter Initials / Date / Time upon completion)
			______ / ______ / ______

18.3 Key Business Process: ______________________________

Start Time from BCP Activation (Days, Hours)	BCP Task	Assigned To	Completion Verification (enter Initials / Date / Time upon completion)
			________ / ________ / ________

<Please copy this section for each key business process.>

Section V - BCP Plan Deactivation and Restoration of Normal Operations

Once the emergency situation has passed or is no longer an impact to the business, it is important to know what to do to get back to normal business operations and when to start doing this. This information needs to be defined before an emergency happens.

19. BCP Plan Deactivation

In this section, document the BCP De-Activation Trigger for each Key Business Process; what would need to happen in order to stop following the BCP plan and go back to conducting business like you did before the emergency (Ex. *Utilities at primary work location have been restored*). Be sure to document who is responsible for deactivating the BCP and initiating the restoration plan. Also be sure to document the affected Contact Tree (i.e. who needs to be informed of the deactivation of the BCP and initiation of the restoration).

BCP ID	Key Business Process	BCP De-Activation Trigger	Responsible Individual(s) for De-Activating BCP and initiating Restoration Plan	Restoration Plan ID (refer to section 20)	BCP Contact Tree* (refer to sections 3 and 4)
19.1					
19.2					
19.3					
19.4					
19.5	<please expand this list to include information for all key business processes>				

20. Return To Normal Operations

In this section, define the tasks needed to move from operating under the BCP plan to restoring normal operations. For each task, document who is assigned the responsibility for completing the task. The tables below are set up to document the tasks, who is responsible and a section for verification that the task was completed. As an alternative to these tables, the restoration plan can be a separate document attached to this document as an Appendix.

20.1 Key Business Process: ______________________

Restoration Task	Assigned To	Completion Verification (enter Initials / Date / Time upon completion)
		______ / ______ / ______

20.2 Key Business Process: ______________________

Restoration Task	Assigned To	Completion Verification (enter Initials / Date / Time upon completion)
		______ / ______ / ______

20.3 Key Business Process: ______________________

Restoration Task	Assigned To	Completion Verification (enter Initials / Date / Time upon completion)
		______ / ______ / ______

<Please copy this section for each key business process.>

Section VI - BCP Plan Testing

To ensure that a BCP plan will be effective in the event of an emergency, it is important to test it out. Just like a fire drill tests out how good your plan is for evacuating the area in an emergency and how well people are trained and prepared to respond to an evacuation, you need to know how well your BCP plan is set up and how well people can follow it. Testing will identify any areas that need attention or revising. Ideally, the test should be documented in a report which is attached to the BCP plan to document what was done, what the results were and any follow up actions that need to be taken to ensure effectivity of the BCP plan.

BCP Plan Test Date	Completed By (Printed Name/Signature/Date)
	____ / ____ / ____

Section VII - BCP Plan Maintenance

It is important to keep a BCP plan current and up to date in order for it to be useful when you really need it. This means that the plan must be reviewed and information updated on a regular or as-needed basis. This can either be done every year or every 6 months as it makes sense, or it can be every time there is a change that impacts on the information contained in the BCP plan.

When the plan is reviewed, indicate the date the plan was reviewed and what changes, if any, were made. If the plan is reviewed on a regular time basis, then it is possible that the review will find that all of the information is current and correct and thus no changes are necessary. It is still important to indicate the date and completion of this review to specify that the plan is still current.

BCP Plan Review Date	Updates / Changes Made, if any. (brief description of sections updated)	Completed By (Printed Name/Signature/Date)
		_______ / _______ / _______

Section VIII - Appendices / Attachments

Appendices or attachments are supporting documents that are attached to the BCP plan document. It is important to have these documents on hand in the event of an emergency. By attaching copies of them to the BCP plan, you will have these all in once place in case you need them and cannot access your regular work location.

Appendix ___ - <insert title here>

Examples:

- Map to alternate locations
- List of suppliers
- Contact list for temporary resources
- Customer list
- Instructions on how to set up a remote office network
- List of IDs and passwords for online systems
- BCP Plan test reports

PART 2 – Instructions for Setting Up and Maintaining the BCP Plan.

Section I - Plan Overview and Contact Information

1. Plan Summary

The purpose of the Plan Summary is to identify the name of the company or organization for which this plan applies and to explain the purpose of the plan (what it is designed to accomplish). The text in the template may be left as it is, but you will need to replace <Company Name> with the actual name of the company or organization. In all of the highlighted sections of the template, any organization-specific information is indicated as <text>. You will need to replace this text with real information as it applies to your company or organization for which you are implementing the BCP plan. This includes any headers or footers in the document.

The plan summary in the template summarizes all of the key areas that the plan addresses. This is important for presenting this plan to auditors and regulatory authorities. They will want documentation to provide evidence that you have covered all the important bases in your business. Critical elements to consider in implementing a BCP plan include the items listed below, as applicable to your particular business and situation:

- Data back-up and recovery (hard copy and electronic).
- Continuation of all mission critical business operations.
- Operational assessments (impacts to business resulting from interruptions to key business operations).
- Alternate communications between customers and the business.
- Alternate communications between the business management and its employees, contractors and suppliers.
- Alternate physical location of employees.
- Regulatory reporting.

2. Plan Approval

All BCP plans must be approved by at least one individual in a senior company management role. This provides evidence that the business is committed to enacting and enforcing the BCP plan (it is just not a stack of papers that they asked someone to write up and print out to satisfy the auditors).

3. BCP Leadership Roles and Responsibilities

Every BCP plan must have specific roles and responsibilities assigned and agreed to before the plan is approved. In the event of an emergency, each person must already be aware of what their responsibility when the BCP plan is activated or the plan will fail. Once an emergency occurs, it is too late to try to figure this out.

BCP teams are typically broken out by functional area (i.e. Sales, IT, Production) or by site (i.e. Building 1, Building 2, or First Floor, Second Floor, or Springfield office St. Louis Office…). This depends on the size and structure of your organization. If you have a small organization, everyone can be on one team. The goal is to keep the tasks manageable such that the team can be activated and perform their duties quickly and effectively in the event of an emergency. If the team is too big or too spread out, this may lead to time delays due to communication lag time and taking on tasks that are too big. Ultimately it is up to the owner of the BCP to decide how to structure the teams. These are just points to consider.

The key responsibilities to consider in setting up the teams are the leadership assignments and responsibilities, and identifying team membership for each person in the organization. Before an emergency occurs, each person must know who they are to report to and each leader must know who they are responsible for.

The key leadership responsibilities must include:

- Who is responsible for activating and deactivating the BCP plan. In the event of an emergency, there must be one primary person who has this responsibility. Activating the BCP plan has significant financial and operation impacts to the business and only a person authorized to make this call should activate the plan. This person must be in a position of authority to have the business information they need in order to be able to make an informed decision to activate the plan. They must also be in a position to quickly receive notification of any events which could trigger the BCP (i.e. Fire, Massive power outage, earthquake, chemical spill, etc.).
- Who is responsible to provide leadership and decision making authority to the BCP team upon activation of the BCP plan, until plan is deactivated. This can be the same person who activates the plan and leads the team. This is up to you to decide, but you must spell this out or else there will be mass confusion when an emergency strikes. People need to know clearly who to take direction from. There needs to be one identified authority to do this.
- Who will manage external communications between the company and outside parties and internal communications between the BCP team and company personnel. This is the person who receives and sends information to and from others inside and outside the company. This includes sending/receiving information to/from local emergency management personnel/police/fire department/etc. and for overseeing communications of BCP/emergency information to/from company personnel. In the event of an emergency, there must be a clear line of communication and everyone needs to know where to go to get accurate and clear information. Without planning this critical piece of the BCP plan, the resulting frenzy of informal communications will cause massive confusion and panic. This needs to be avoided or else the plan will fail.
- Who is responsible for the BCP plan itself; the BCP plan administrator. This is the person who owns and manages the BCP plan to ensure it is current and accessible in the event of an emergency. This is a key role which needs to be taken seriously as a primary responsibility to get this plan in place and keep it up to date. Ensure review and testing is performed and documented on a regular basis. This means the plan is checked to make sure the information is still current and accurate. Also make sure that the plan is tested out to make sure it will work and to find out where revisions or additions need to be made to fill any gaps uncovered during testing. This is not just something do when time permits, or else it will not get done or it will get out of date and then when an emergency strikes, you will be out of luck trying to scramble to find out what information is current

and what is not. Time is money and the time you put in up front to get this plan in shape will pay off later on when you need it to be correct the first time you activate it.

- Who is in charge of each sub team (Functional Lead). These are the people who are responsible for leading each of the BCP teams as described above.
- VERY IMPORTANT: For each of these roles described above there must be a primary and a backup person identified. In the event that the primary person is not available or unable to perform their role, the backup person must take that role. The backup person must be made aware that they have been designated as the backup person for their specific role before the emergency happens. They must have access to the same information as the primary person has and they must take part in all BCP training and testing exercises so that if needed in an emergency they can step into the primary role without any lag time.

4. Internal Contact Information

In the event of an emergency or disaster, it is critical to have contact information for all personnel to be able to quickly share important information as it becomes available and to quickly account for all personnel. The most useful format for this information, in the event of an emergency, is to set it up as one or multiple contact trees (depending on how big your organization is).

The contact tree is a structured hierarchical format with multiple levels which look like a pyramid if you draw them in a diagram (see example below).

Example Contact Tree Diagram

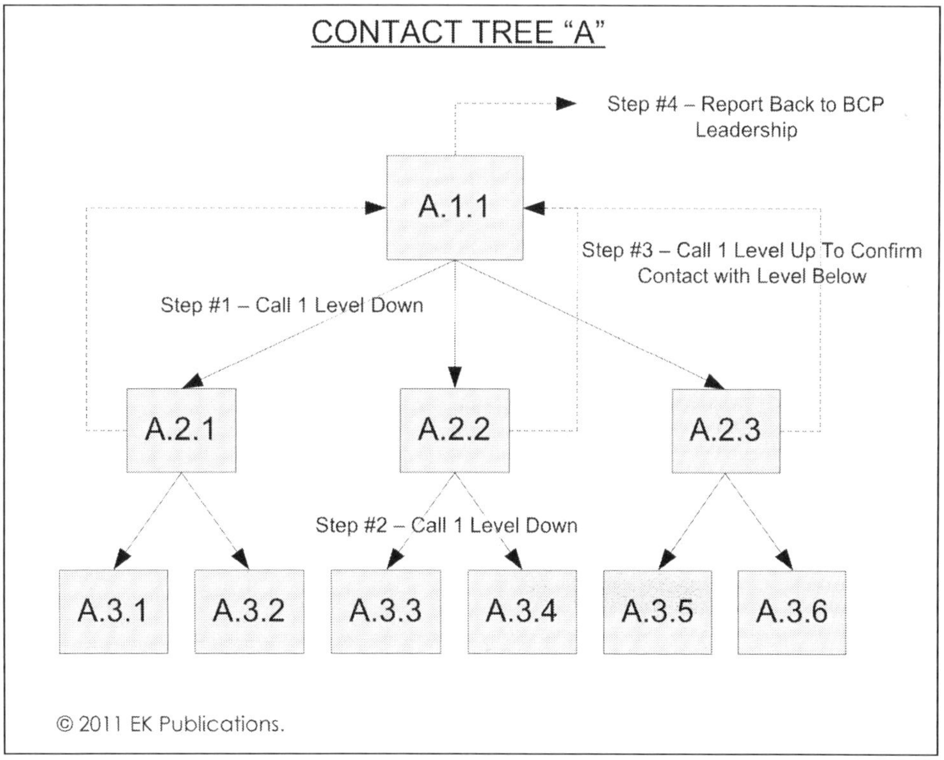

The critical pieces are:

- Each person is assigned a specific box in the contact tree.

- There are defined levels to ensure clear direction of communication.
- The tree is designed to map a logical grouping of individuals (i.e. by function, by department, by site).
- The tree is designed to distribute the tasks of making contact (i.e. no one person does all the work). Ideally each person should have 1-5 contacts to make.
- Each box is clearly associated with levels above and below.
- Each person knows who will contact them, and in turn who they will contact.
- Multiple means of contact are identified for each person (daytime and off-hour phone, cell phone, email, etc.) so that if one means of contact does not work another can be attempted.
- Special skills are identified which could be vitally important in an emergency situation (i.e. who is a doctor, paramedic, fire fighter, etc. who can be asked to help tend to any urgent situations).
- There is a confirmation that loops back up the tree to ensure that all the people in the tree are accounted for, or to alert the top level that people cannot be accounted for. It is not enough just to call down the tree and presume that everyone has gotten the message; you need to know for sure that everyone has been reached.

This is one of the most important pieces of the BCP. Without basic communications, the rest of the plan will fail. Be sure this section can easily be printed out and provided to all personnel in the tree. In the event of an emergency, each person needs to access this plan quickly and must not be dependent on any electronic or online systems which may not be functional. An electronic copy is not sufficient if the power is out or you cannot get to your computer.

This is how the contact tree works -

In the event of an emergency, the BCP leadership contacts the person at the top of the tree (Level 1). The Level 1 person then contacts the assigned people in the tree one level down (Level 2). The Level 2 people then contact the people assigned to them one level down (Level 3). Once the Level 2 people have contacted (or attempted to contact) all the Level 3 people assigned to them, they then contact the Level 1 person above them to report the status (All People Accounted For or People Not Accounted For) and any questions or information to be shared. This provides confirmation that the contact tree has reached all the affected individuals and that everyone is accounted for. The Level 1 person at the top of the tree is then responsible for reporting back to the BCP Leadership the status of people in their contact tree and conveying any information they need to share.

The contact tree can be set up various ways depending on what makes sense to your organization such as supervisor with associated personnel below, or it make more sense to have a level 1 site person with assigned people at the same site below (for organizations where reporting structures span multiple locations).

The contact tree can be expanded (more levels added) for larger groups. Multiple contact trees can be set up for multiple groups, functions, or locations.

The table provided in the template is set up in a contact tree structure to identify each person within a contact tree and to provide instructions on what to do in the event the contact tree is activated. Information about each person should include how to reach them through multiple means (phone, email, text, etc) and at different times (work hours, after work hours, etc). It is also very useful to note any special skills or training each person has (i.e. paramedic, firefighter, doctor, nurse, etc.) in the event their services are needed in an emergency.

5. External Contact Information (Customers)

This section documents contact information for your customers. In the event of an emergency, you must be able to continue to communicate with your customers in order for the business to continue. To be prepared for this, you need to have a list of your customers and key information. This should include at a minimum their name and contact information and information about any orders/contracts they have with your business. This list can be documented in the table included in the template or it can be a report from one of your systems. Depending on how big your business is, you may choose to have only the very top customers' information documented in portable format (i.e. printed out so you can take it with you) and have the rest of the customers' information saved in a computerized system (which has a backup/disaster recovery plan in place to ensure you can access a copy of this data in the event that the primary system is not functional). The bottom line here is that in the event you need to contact a customer or client in an emergency situation, you can still do it.

6. External Contact Information (Suppliers and Business Service Providers)

This section documents information you will need to keep in contact with your service providers and suppliers. These include suppliers of services or materials needed by your business, legal services, business services, etc. These contacts will be especially important in the event of an emergency (i.e. contacting the insurance company in the event of disaster or notifying your suppliers of alternate work locations) so it is important to have the information easily accessible. It is best to have this in one list, which includes the names and contact information as well as account or policy numbers you will need to identify yourself. It is not advisable to rely on multiple sources for this information of old documents (i.e. bills, brochures, quotes) as these will take time to find or may not be accessible at all during an emergency. One option would be to scan these documents and create a single electronic document which you can copy (keep a backup in an alternate location) and print out to keep with you. Another would be if you have this information in a database or computer system, run a report which summarizes the information you need and print this out to keep with you.

7. Utilities and Facilities Services

This section is very similar to section 6 but focuses on utility companies and facilities services such as leasing companies. It is very important to have this information easily at hand in the event of an emergency where you might need to contact the utility company to shut off service or report an outage etc.

8. Financial Services Contact Information

This section documents the information you will need to communicate with financial services (i.e. Bank, payroll and benefits services, etc.) and access your accounts. In the event of an emergency or disruption to business requiring relocation, it is imperative to maintain contact with financial institutions to enable continuation of payments to employees, suppliers and service providers, as well as processing payments from customers.

The list must contain the names and contact information for all of your financial services so that you have this in one place easily accessible in the event of an emergency. Sensitive information such as account numbers, log on IDs, PINs and passwords are important to be able to access your accounts but you probably do not want these printed out on a list that someone can copy and use to get into your accounts. So the best way to manage this is to designate key individuals in your organization who have these pieces of information and are responsible for managing them and keeping them confidential. For purposes of the BCP plan you then need to identify the individuals (make sure you have a primary and a backup person identified) who is responsible for managing the account and password for each financial service. In the event of an emergency, this person would be contacted and asked to access the account and perform whatever management needs them to perform to keep the business operating. This is another reason why the contact tree is so vitally important because you want to be sure you can contact these people.

9. Regulatory Agency Contact Information

This section documents contact information for regulatory agencies and authorities such as IRS, Tax departments, local authorities and other regulatory bodies depending on your business (i.e. .FDA, EPA, DEP, Customs, Licensing agencies, etc. as applicable). It is critical to maintain communications with regulatory agencies during emergencies to enable reporting of business status affecting required regulatory filings or reporting time sensitive business critical information to regulatory agencies. You don't want to add to your troubles by getting behind on renewing your license to manufacture your product or paying your taxes because you cannot locate the information you need because it is on a computer which you lost in a flood.

10. Critical Records and Systems

This section identifies your business critical records and systems. This includes paper files and computer systems which contain business critical data (such as orders, financial transactions, product specifications, product manuals, software source code, etc.). For all of these records and systems you need to have a back up system and/or location identified and documented in a list that is easily accessible in the event of an emergency. Backup locations and systems include paper copies, microfilm or scanned copies of paper documents stored at an alternate location, and system data and software backed up on DVD's or tapes etc. and stored at an alternate location.

It is not necessary to document system IDs and passwords here (unless you want to) but at a minimum you need to identify who has the necessary system IDs and passwords, pin, keys, combinations, etc. to be able to access the records/systems. Be sure to identify a primary and backup person here.

Ideally backup copies of systems and data should be copied onto multiple media. Most modern libraries that backup irreplaceable data use multiple media such as optical media (CDs, DVDs) plus magnetic media (Tapes, Hard Drives) and solid state media (Memory Stick, Jump/Flash/Thumb Drive). With the emergence of cloud computing, the new trend is towards online storage which is accessible anywhere on the Internet. Services of this type are provided by companies like Amazon or Online Mailbox. The important piece to consider is that you need to be able to access important information and you should not rely too heavily on just one media because there are multiple causes of failure for each type (i.e. Tapes can be demagnetized, CDs can warp if exposed to heat, Jump drives will fail if they get wet, etc.).

For all of the storage media options and for copies of paper document, at least one copy of these needs to be kept in an alternate location away from your primary work location just in case you cannot access your primary work location. Having your critical files in a fireproof safe which is in your office which you cannot get to because of an evacuation due to a massive chemical spill is not going to help you. You need a copy that is either with you at all times or in a location away from the primary location. The general rule is that the backup location should be approx 20-50 miles away from the primary location (close enough so that you can get to it easily but far enough away so that it is not affected by the disaster). The back up location should ideally be protected from theft, fire and water damage and should ideally be climate controlled (temperature and humidity are controlled to minimize damage to your records and media). There are companies such as Iron Mountain who provide this type of service.

11. Backup Locations

This section documents the backup work locations identified which are to be used in the event of an emergency which prevents work from happening at the primary location. Some tasks can be performed remotely such as in a home office (i.e. Sales functions or technical writing) but others will require backup locations (i.e. manufacturing, laboratory work). Identification of backup locations and agreement on the terms of using the backup locations need to be defined and documented before the BCP plan is finalized.

These considerations include:

- Space requirements vs. availability
- Qualification/certifications required to perform the work at the backup location
- POs /payments/contracts or financial agreements required
- Time needed between notification and occupying the backup location
- Using internal personnel at backup location or outsourcing the operation entirely

The purpose of this section is to ensure continuation of critical business functions during activation of the BCP plan. During an emergency you want to know who to contact to get business operations moving quickly. You cannot afford to be researching options for resuming business operations. Backup/contingency locations must be identified for all essential operations which cannot be performed remotely (at home). Backup sites should be far enough away from the primary work site to reduce the chances that the disaster will affect access to the backup site, but not so far as to prevent timely access by personnel. A general rule is for the backup site to be 20-50 miles from the primary site. Backup site information in the BCP plan must include directions instructing personnel how to get to the location.

Directions and Maps to all backup locations need to be included in the Appendix Section of the BCP to be able to easily provide copies to all personnel who need to report to the backup locations.

12. Backup Service Provider and Supplier Information

This section documents a list of your emergency/backup suppliers. These are the people or companies you would go to in the event that you cannot use your normal or primary suppliers either because you have moved work locations during an emergency or because there is an emergency at the primary supplier which makes them unable to help you.

In the event of an emergency or disruption to primary business services and suppliers, a list of back up services and suppliers must be maintained for quick reference to enable restoration of essential business functions in a timely manner. This list should include backup suppliers for essential functions necessary for the business to operate, such as computer services, shipping, order processing, etc.

13. Emergency Services Contact Information

This section is a basic list of emergency services such as police, fire, etc. In any emergency situation, it is essential to know who to contact. This information must be on hand for easy availability. List all critical emergency contact information in the table below so you have it all in once place for easy access.

Section II - Business Risk Assessment and Impact Evaluation

This section of the BCP plan documents the risk associated with the disruption of any key business operation or process. Risk is defined as the combination of how likely a key operation will be disrupted, how much time before the business experiences the negative impact of losing the operation, and how much this disruption will hurt the business performance. (*Ex. The manufacturing plant can be shutdown for several months in the event that the river running through the property overflows and floods the building. Some flooding has occurred once every 3 years with major flooding occurring on average every 25 years. This will result in loss of sales within 2 days*).

The information documented in this section is used to justify the prioritization of the BCP plan activities and justify the money allocated to ensuring the continued operations during an emergency or disaster. (*Ex. A backup manufacturing site is qualified and PO-##### is issued to Company Z to enable start up of production within 1 day in the event of an emergency at the primary manufacturing facility. A new manufacturing site has been identified and a project initiated to relocate primary manufacturing to this site by 15-JUN-20##*).

The more these figures can be quantified, the better will be the BCP decision-making.
To assist in this evaluation, use the scale indicated in the chart provided in the template.

Once this section is complete, this information will provide a roadmap for prioritizing the BCP tasks. The result will be a list of key business operations/processes with risk levels assigned. The operations/processes with the lowest numbers will have the highest risk level. The high risk items are the ones which should receive the most attention when designing and implementing the BCP plan. These are the items which will hurt the business the most if they are not adequately addressed.

14A. Business Process Risk Evaluation

This section documents the risk associated with each business operation or process. In order to create a BCP plan that is effective and manageable it is important to understand the relative risk associated with an interruption to each critical business operation so that the BCP plan can address restoring these operations in a prioritized ordered approach. It is impossible to go from a disaster which interrupts your business to bringing everything back online all at once. There is so much to manage and many limitations during emergencies that you need to be very clear before the emergency happens what your priorities are. If you lost your primary work site, what operation/function would you need to bring back first? Second? Third? Etc.

The critical elements of completing this section are:

- Identifying the <u>key business operations and processes</u>. These are operations or process which must be working in order for the business to continue. Examples are customer service, processing orders, shipping product, etc.
- Identifying the <u>likelihood of disruption</u> (chances that something unexpected could happen which disrupts the key business operation or process). **SCALE: 1 – 10** (1 = Very <u>High</u> Likelihood of Unplanned Disruption, 10 = Very <u>Low</u> Likelihood of Unplanned Disruption)
- Identifying <u>how significant the effect is of losing this operation or process.</u> (i.e. people at risk, profits lost, customers lost) **SCALE: 1 – 10** (1 = Very <u>High</u> Significance To The Business, 10 = Very <u>Low</u> Significance To The Business).

- Identifying how long this operation can be disrupted before it begins to negatively impact the business. **Number of Days** without this operation before business would experience any negative impact to operations and/or sales.

14B. Business Process Disruption Impact Evaluation

In this section, the impact to the business of disruptions must be evaluated for each critical business operation identified in Section 14A. The object here is to define how much of an impact will disruption of each critical business process have on the business.

This must be defined in terms of:

- Available work-arounds to temporarily replace the process *(ex. paper records can be processed manually in the event that a key computer system is unavailable, process can be outsourced to Company X).*
  - A workaround is a manual process or other way to accomplish what needs to be done in the event that the primary process is unavailable. For example, you can take orders on paper forms if the computer system is down or you can manually produce a product if the automated production line is down, etc.
  - If there is a workaround, there will be less impact to the business as the BCP plan should specify that the workaround is activated in the event that the operation/process is disrupted. The BCP plan should include instructions on how to perform the workaround.
  - If there is not a workaround, the impact to the business will be greater because when the primary operation/process is interrupted your business will feel the effects (i.e. sales lost, no more orders taken, etc.)

- How long can the business continue to operate until the process must be restored? *(ex. computer system must be back online within 3 months or costs and loss of efficiency will reach unsustainable levels).*
  - This is not the same as the question in section 14A (how long before the business begins to feel the effect of a disruption to the operation). This is how long before you decide the process must be restored. This is different because due to priorities and economic/resource limitations, you need to make the proactive decision to wait a certain period of time before taking action to restore this operation due to the cost and resource considerations this will have vs. the risk of waiting. The information documented in Section 14A (how long until the business begins to feel the negative effect to the operation) is a key piece of information needed to make an informed decision on how long to wait until restoring the operation. Other operations may take priority.
  - This needs to be decided and documented before an emergency happens.

- What resources are needed for the work-around and for restoration of the process?
  - In order to effectively plan for restoring a critical business operation, you need to know what resources (people, time, materials, supplies, money) are needed.
  - This information needs to be documented in advance of an emergency so that when you need to activate the BCP plan you know right away what you need. Even better, you can put together some of these supplies and materials ahead of time and have it ready just in case.

List the Key Business Processes in order of BCP Priority (lowest number first) using the information identified in the table in question 14A. The objective here is to identify which processes require the most immediate attention in the event of a disruption and what is required to address the disruption.

Section III - Pre-Emergency Planning

15. Risk Mitigation

The best way to prevent getting caught in a disaster is to take proactive action and see what you can do to prevent a potential disaster or at least minimize the impact. For example if you identified that storing flammable materials, such as spray paint and paint thinner, near your hot water heater (which is heated by gas and has an open flame) could result in starting a fire, your mitigation action would be to move the flammable materials to a safe location to reduce the risk of a fire starting.

This same logical approach needs to be taken with your business. If you find that you have data that is not backed up or if you have computer systems which are vulnerable to viruses or unauthorized access, you need to either determine that the risk is acceptable or take action now to reduce your identified risk as best you can. Not every risk must be mitigated. Some are not possible to mitigate so they need to be planned for in your BCP, some are too costly to mitigate, and some may not carry a high enough risk for you to spend resources on mitigating. Only you can determine this, but you need to go through this exercise so you can make informed decisions and plan accordingly.

16. Emergency Materials and Supplies

To prepare in advance for an emergency, it is vital to have emergency supplies on hand and easily accessible to the people who need them. This should include critical emergency materials and supplies that need to be available in the event of an emergency or unplanned business interruption. This includes small disposable supplies (i.e. pens, paper, mailing supplies, etc.), office equipment (computers, printers, phones, etc.), production equipment (i.e. machinery needed to service customers, refrigerator, freezer, generator, batteries, tools, etc.) as well as raw materials and production supplies (i.e. paper, paint, lawn fertilizer, baking ingredients, gasoline, propane, etc.).

In the previous section, you identified what supplies are needed to restore each business operation. Now that you know what you need, you must decide how much of this you will acquire and keep on hand before an emergency happens. This is determined based on the cost of keeping in emergency supplies on hand versus risk of not having them. Only you can determine this.

Some supplies can be easily obtained if needed in a hurry but others are more difficult. During a flood, I know from experience it is impossible to find sump pumps or generators. Even the large suppliers like Home Depot and Lowes ran out of sump pumps and generators and did not expect more for another week, and they had a waiting list of people wanting them. Meanwhile the water level was rising in my newly finished basement. Consider this when completing this section.

Section IV - BCP Plan Activation and Implementation

In this section, identify the triggers for activating the BCP plan (what events must occur in order to activate the BCP plan), who is involved in activating the BCP plan, and what are the steps to activate and implement the plan. All key business processes identified in Section 14A should be listed in Section 16 in order of BCP priority.

Start times should be specified based on impact evaluation information captured in section 14B (*ex. if production line downtime causes loss of customer orders within 2 days, then trigger is that production line is down for more than 1 day and target start time for production operation BCP should be specified to have a backup process in place within 2 days*).

17. Plan Activation

This is the section which needs to contain the specific trigger for activating the BCP plan.
It must clearly identify:

- What needs to happen in order for the BCP plan to be activated?
- Who is responsible for activating the BCP plan (the primary and backup person)?
- Which BCP plan (the instructions for addressing the disaster) are to be followed?
- Which contact tree needs to be activated to notify the proper individuals of BCP activation?

It is disruptive and can be expensive to activate a BCP plan, so you need to be clear up front what situations warrant the activation of the BCP. Activating a BCP is not appropriate for every interruption. For example, if there are downed wires that knock out power for 6 hours to your primary manufacturing site, this may or may not be enough to activate the BCP. You need to decide this based on the evaluation performed following the template. This depends on how long you can be without the operation versus how long it would take to get the backup operation going and how much it would cost. You may decide you can afford to wait for up to 2 days without power before activating the BCP. But you need to do this evaluation and document the trigger up front before the emergency happens. If you know you can wait 2 days, then the appropriate action is to start counting the time from when the power is interrupted and at 2 days you need to activate the BCP plan if the power is not coming back on.

Having a plan is not enough; you need to be sure when to use it in order for it to be effective otherwise you could be wasting time, effort, and money by needlessly panicking for no reason.

18. Plan Implementation

This section provides the instructions for carrying out the BCP plan. It needs to be very clear and easy to follow. It is highly recommended that one plan is written for each critical business process identified in the template. The plans should be written clearly in stepwise format (i.e. Do Step 1, then Step 2, then Step 3). For each step there needs to be a person assigned to do this step, so no one assumes someone else is taking care of something (and then nobody does it and it doesn't get done!). It is also good practice to have a signoff for each step so you have written confirmation that a step has been completed. This is especially helpful in reviewing BCP implementations and tests to see what worked well and what needs improvement. This is also very helpful for BCP plans which involve handoffs (i.e. person A completes Step 1 and then Person B completes Step 2). The tables in the template are set up to document the tasks, who is responsible and a section for verification that the task was completed. As an alternative to these tables, the implementation plan can be a separate document attached to the BCP Plan as an Appendix.

Section V - BCP Plan Deactivation and Restoration of Normal Operations

This section defines what happens once the emergency situation has passed or is no longer an impact to the business. The objective here is to return to normal business operations. First you will need to define what must happen in order for you to begin the process of returning to normal business operations or to a longer term solution (the BCP deactivation trigger), and then you need to define what the instructions/steps are to deactivate the BCP and return to normal operations (deactivation implementation/return to normal operations).

This section is set up just like the BCP Activation and Implementation section above but it is to provide instructions on the opposite process. Your objective is to be able to get back to a normal mode of business and you need to know when and how to deactivate BCPs as soon as it is appropriate.

19. BCP Plan Deactivation

In this section, document the BCP De-Activation Trigger for each Key Business Process; what would need to happen in order to stop following the BCP plan and go back to conducting business like you did before the emergency (Ex. *Utilities at primary work location have been restored*). Be sure to document who is responsible for deactivating the BCP and initiating the restoration plan. Also be sure to document the affected Contact Tree (i.e. who needs to be informed of the deactivation of the BCP and initiation of the restoration).

20. Return To Normal Operations

In this section, define the tasks needed to move from operating under the BCP plan to restoring normal operations. For each task, document who is assigned the responsibility for completing the task. The tables in the template are set up to document the tasks, who is responsible and a section for verification that the task was completed. As an alternative to the tables, the restoration plan can be a separate document attached to the BCP Plan as an Appendix.

Section VI - BCP Plan Testing

To ensure that a BCP plan will be effective in the event of an emergency, it is important to test it out. Just like a fire drill tests out how good your plan is for evacuating the area in an emergency and how well people are trained and prepared to respond to an evacuation, you need to know how well your BCP plan is set up and how well people can follow it. Testing will identify any areas that need attention or revising.

The degree to which the testing is conducted is based on what you feel is appropriate. Obviously you will not set your office on fire to test the BCP, but many organizations do conduct emergency backup testing of computer systems (disaster recovery testing to demonstrate that the system can be brought up at the backup site and it is still accessible and still works and provides the data needed to operate the business). Other tests include simulations to varying degrees. Some examples are having everyone work remotely for one day to verify how well the networks and systems operate and note any problems. Or do some laboratory testing or manufacturing at the back up site to verify how well and how quickly and how productive the backup plan works.

The objectives of this section are:

- To demonstrate that the plan does work.
- To provide people an opportunity to be trained in the BCP so they ready for a real emergency.
- To identify any problems, issues or areas of improvement that need to be addressed in the plan.

As a result of testing you may find out that the backup server needs more disk space, or the data backups need to happen more frequently or people working remotely may not be able to access necessary systems because of firewall issues, or your backup manufacturing site does not have a loading dock so all you materials needs to be carted using fork trucks. The only way to realize these potential problems is through testing.

Make you sure you at least test the contact tree on a regular basis. You should be able to contact all the people in the tree and get confirmation back. Note how long this takes and any difficulties you encounter. These need to be addressed as one of the most basic and fundamental failures in a BCP plan is not being able to get in touch with everyone involved. If people have important roles that are critical, you might need to get them a better cell phone or faster internet service so you can communicate with them in an emergency. During the Y2K planning, key individuals in the company I worked for were given satellite phones in the event that the cell phone network went down. You need to decide what is appropriate.

Testing should be performed on a regular basis which you must determine based on what you think is appropriate. Some organizations perform yearly testing, or partial testing one year and full testing the next. When significant changes happen (you use a new back up site for your servers or take on a new manufacturing process) you would probably want to test the BCP to ensure these work as you intend.

Ideally, the test should be documented in a report which is attached to the BCP plan to document what was done, what the results were and any follow up actions that need to be taken to ensure effectivity of the BCP plan.

Section VII - BCP Plan Maintenance

It is important to keep a BCP plan current and up to date in order for it to be useful when you really need it. This means that the plan must be reviewed and information updated on a regular or as-needed basis. This can either be done every year or every 6 months as it makes sense, or it can be every time there is a change that impacts on the information contained in the BCP plan.

The review should look at all the critical processes and make sure none are missing or have changed significantly such that the BCP plan no longer is adequate or accurate. You also need to carefully review contact trees to make sure the right people are in the right places in the tree. And you need to ask people to confirm that their contact information is still correct.

When the plan is reviewed, indicate the date the plan was reviewed and what changes, if any, were made. If the plan is reviewed on a regular time basis, then it is possible that the review will find that all of the information is current and correct and thus no changes are necessary. It is still important to indicate the date and completion of this review to specify that the plan is still current.

Section VIII - Appendices / Attachments

Appendices or attachments are supporting documents that are attached to the BCP plan document. It is important to have these documents on hand in the event of an emergency. By attaching copies of them to the BCP plan, you will have these all in once place in case you need them and cannot access your regular work location.

Examples:

- Map to alternate locations
- List of suppliers
- Contact list for temporary resources
- Customer list
- Instructions on how to set up a remote office network
- List of IDs and passwords for online systems
- BCP Plan test reports

PART 3 - EXAMPLE: Business Continuity Plan (BCP) for EK Catering Service, LLC

This document provides an example of how a completed BCP plan looks. It is made up for a fictitious company with completely made up names, addresses and emails. The purpose is to illustrate the type of information, organization and level of detail to be documented in a BCP. The instructions will help to explain how to use the BCP template but an example to illustrate the finished product is worth a lot more.

Section I - Plan Overview and Contact Information

1. Plan Summary

This Business Continuity Plan (BCP) has been developed to ensure EK Catering Service, LLC, located at 105 Main St, Downtown, MA 01234, is prepared to respond to unplanned business interruptions that may affect critical business processes. An unplanned business interruption is defined as an event (such as an emergency or disaster) that causes an unanticipated disruption to the operation of the business and/or service to customers. The objective of this plan is to minimize the impact of unplanned interruptions and provide guidance for continuing operations during an unplanned event and resuming normal operations once the event is resolved.

2. Plan Approval

All BCP plans must be approved by at least one individual in a senior company management role.

	Printed Name	Title	Signature	Date
2.1 Prepared By:	Janet Kline	Customer Service Manager	Janet Kline	September 23, 2011
2.2 Approved By:	Susan Smith	Owner, President	S. Smith	September 23, 2011
2.3 Approved By:	Steven Brown	Co-Owner, Vice President	Steve Brown	September 23, 2011

3. BCP Leadership Roles and Responsibilities

BCP leadership roles and responsibilities must be identified and defined before an emergency occurs in order for the BCP plan to be effective.

BCP ID	Role	BCP Responsibility	Name
3.1	Company Management / Site Management (Primary)	Activate BCP plan upon notification of emergency, disaster or unplanned event causing interruption to business operations. Deactivate the BCP plan when conditions allow for transition back to normal operations.	Susan Smith
3.2	Company Management / Site Management (Backup)	Backup person who can activate the BCP in the event that the primary person responsible for BCP activation is unavailable.	Steve Brown
3.3	BCP Team Leader (Primary)	Provide leadership and decision making authority to BCP team upon activation of the BCP plan, until plan is deactivated.	Stan Wilson
3.4	BCP Team Leader (Backup)	Backup person to serve as BCP leader in the event that primary BCP leader is unavailable.	Janet Kline

BCP ID	Role	BCP Responsibility	Name
3.5	BCP Communications Leader (Primary)	Manage external communications between the company and outside parties and internal communications between the BCP team and company personnel.	Stan Wilson
3.6	BCP Communications Leader (Backup)	Backup person to serve as Communications Leader in the event that primary Communications Leader is unavailable.	Sally Jones
3.7	Plan Administrator (Primary)	Own and mange the BCP plan to ensure it is current and accessible in the event of an emergency. Ensure review and testing is performed and documented on a regular basis.	Janet Kline
3.8	Plan Administrator (Backup)	Backup person to serve as Plan Administrator in the event that primary Plan Administrator is unavailable.	Stan Wilson
3.9	Functional Lead (Primary) Customer Service	Functional leader/manager in charge of Customer Service.	Janet Kline

BCP ID	Role	BCP Responsibility	Name
3.10	Functional Lead (Backup) Customer Service	Backup person to serve as Functional leader/manager for Customer Service in the event that primary Functional leader/manager is unavailable.	Stan Wilson
3.11	Functional Lead (Primary) Kitchen/Food Preparation	Functional leader/manager in charge of Kitchen operations.	Phil Smith
3.12	Functional Lead (Backup) Kitchen/Food Preparation	Backup person to serve as Functional leader/manager for Kitchen Operations in the event that primary Functional leader/manager is unavailable.	Linda Hudson
3.13	Functional Lead (Primary) Finance/Billing/Orders	Functional leader/manager in charge of Finance/Billing/Orders.	Kate Green
3.14	Functional Lead (Backup) Finance/Billing/Orders	Backup person to serve as Functional leader/manager for Finance/Billing/Orders in the event that primary Functional leader/manager is unavailable.	Sam Lockwood

4. Internal Contact Information

In the event of an emergency or disaster, it is critical to have contact information for all personnel to be able to quickly share important information as it becomes available and to quickly account for all personnel. The most useful format for this information, in the event of an emergency, is to set it up as one or multiple contact trees (depending on how big your organization is).

The contact tree is a structured hierarchical format with multiple levels which look like a pyramid if you draw them in a diagram (see example below).

In the event of an emergency, the BCP leadership contacts the person at the top of the tree (Level 1). The Level 1 person then contacts the assigned people in the tree one level down (Level 2). The Level 2 people then contact the people assigned to them one level down (Level 3). Once the Level 2 people have contacted (or attempted to contact) all the Level 3 people assigned to them, they then contact the Level 1 person above them to report the status (All People Accounted For or People Not Accounted For) and any question or information to be shared. This provides confirmation that the contact tree has reached all the affected individuals and that everyone is accounted for. The Level 1 person at the top of the tree is then responsible for reporting back to the BCP Leadership the status of people in their contact tree and conveying any information they need to share.

The contact tree can be set up various ways depending on what makes sense to your organization such as supervisor with associated personnel below, or it make more sense to have a level 1 site person with assigned people at the same site below (for organizations where reporting structures span multiple locations).

The contact tree can be expanded (more levels added) for larger groups. Multiple contact trees can be set up for multiple groups, functions, or locations.

The table provided below is set up in a contact tree structure to identify each person within a contact tree and to provide instructions on what to do in the event the contact tree is activated. Information about each person should include how to reach them through multiple means (phone, email, text, etc) and at different times (work hours, after work hours, etc). It is also very useful to note any special skills or training each person has (i.e. paramedic, firefighter, doctor, nurse, etc.) in the event their services are needed in an emergency.

Example Contact Tree Diagram

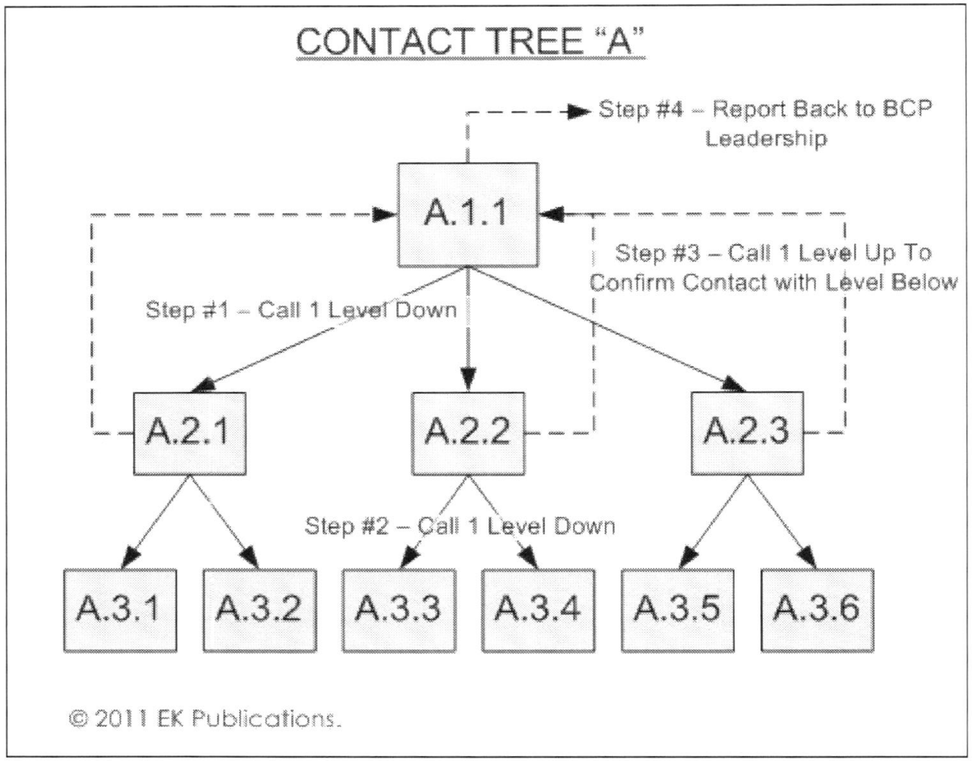

Contact tree ID	Name	Title and Affiliation (Employee, Contractor)	Emergency Skills/Qualifications (ex. EMT, MD, RN)	(W)ork Phone/ (H)ome Phone/ (C)ell Phone	email	Contact tree Instructions
CONTACT TREE A – Home Office - LEVEL 1						
A.1.1	Susan Smith	President	Not Applicable	(W) 212-555-1234 (H) 201-555-4333 (C) 201-444-2231	ssmith@ekcllc.com	- Contact A.2.1___ - Contact A.2.2___
CONTACT TREE A – Home Office - LEVEL 2						
A.2.1	Steven Brown	Vice President	Not Applicable	(W) 212-555-4444 (H) 201-555-3333 (C) 201-444-1313	sbrown@ekcllc.com stevenb@aol.com	- Contact A3.1___ - Confirm contacts complete with A.2.1____
A.2.2	Stan Wilson	Manager	EMT	(W) 212-556-1234 (H) 201-556-4333 (C) 201-445-2231	ssmith@ekcllc.com	- Contact A.3.2___ - Contact A.3.3___ - Confirm contacts complete with A.1.1____
CONTACT TREE A – Home Office - LEVEL 3						
A.3.1	Phil Smith	Manager	Not Applicable	(W) 212-557-1234 (H) 201-557-4333 (C) 201-446-2231	pasmith@ekcllc.com	- Initiate Contact Tree B - Confirm contacts complete with A.2.1_____
A.3.2	Kate Green	Manager	Not Applicable	(W) 212-558-1234 (H) 201-558-4333 (C) 201-447-2231	kgreene@ekcllc.com	- Initiate Contact Tree C - Confirm contacts complete with A.2.2____

Contact tree ID	Name	Title and Affiliation (Employee, Contractor)	Emergency Skills/Qualifications (ex. EMT, MD, RN)	(W)ork Phone/ (H)ome Phone/ (C)ell Phone	email	Contact tree Instructions
A.3.3	Janet Kline	Manager	Not Applicable	(W) 212-559-1234 (H) 201-559-4333 (C) 201-448-2231	jkline2@ekcllc.com	- Initiate Contact Tree D - Confirm contacts complete with A.2.2_____
			CONTACT TREE B – Kitchen Services – Level 1			
B.1.1	Phil Smith	Manager	Not Applicable	(W) 212-557-1234 (H) 201-557-4333 (C) 201-446-2231	pasmith@ekcllc.com	- Contact B.2.1___ - Contact B.2.2___ - Confirm contacts complete with A.2.1____
			CONTACT TREE B – Kitchen Services – Level 2			
B.2.1	Jill Jenkins	Staff Specialist	Not Applicable	(W) 212-560-1234 (H) 201-560-4333 (C) 201-449-2231	jjenkins@ekcllc.com	- Contact B.1.1 as needed.
B.2.2	Linda Hudson	Staff Specialist	Not Applicable	(W) 212-561-1234 (H) 201-562-4333 (C) 201-450-2231	lhudson@ekcllc.com	- Contact B.1.1 as needed.
			CONTACT TREE C – Billing Services – Level 1			
C.1.1	Kate Green	Manager	Not Applicable	(W) 212-558-1234 (H) 201-558-4333 (C) 201-447-2231	kgreene@ekcllc.com	- Contact C.2.1___ - Contact C.2.2___ - Contact C.2.3___ - Confirm contacts complete with A.2.2____
			CONTACT TREE C – Billing Services – Level 2			

Contact tree ID	Name	Title and Affiliation (Employee, Contractor)	Emergency Skills/Qualifications (ex. EMT, MD, RN)	(W)ork Phone/ (H)ome Phone/ (C)ell Phone	email	Contact tree Instructions
C.2.1	Sam Lockwood	Staff Specialist	Fire Fighter	(W) 212-562-1234 (H) 201-563-4333 (C) 201-451-2231	slockwood@ekcllc.com	- Contact C.1.1 as needed.
C.2.2	Bill James	Staff Specialist	Not Applicable	(W) 212-563-1234 (H) 201-564-4333 (C) 201-452-2231	wjames@ekcllc.com	- Contact C.1.1 as needed.
C.2.3	Sue Cross	Staff Specialist	Not Applicable	(W) 212-564-1234 (H) 201-565-4333 (C) 201-453-2231	scross@ekcllc.com	- Contact C.1.1 as needed.
CONTACT TREE D – Customer Service – Level 1						
D.1.1	Janet Kline	Manager	Not Applicable	(W) 212-559-1234 (H) 201-559-4333 (C) 201-448-2231	jkline2@ekcllc.com	- Contact D.2.1___ - Contact D.2.2___ - Confirm contacts complete with A.2.2____
CONTACT TREE D – Customer Service – Level 2						
D.2.1	David Gleason	Staff Specialist	Not Applicable	(W) 212-565-1234 (H) 201-566-4333 (C) 201-454-2231	dgleason@ekcllc.com	- Contact D.1.1 as needed.
D.2.2	Brandon White	Staff Specialist	Not Applicable	(W) 212-566-1234 (H) 201-567-4333 (C) 201-455-2231	bwhite@ekcllc.com	- Contact D.1.1 as needed.

5. External Contact Information (Customers)

A critical piece of information to have available in the event of an emergency causing interruptions to business operations is a customer list to enable timely communications to customers during outages. This list can either be compiled using the table below, or if customer information is maintained in a separate system and a list can be printed or copied and pasted, attach it to this BCP plan as an Appendix and reference it here in place of the table.

BCP ID	Name	Title	Company Name and Address	(W)ork Phone/ (H)ome Phone/ (C)ell Phone	Web address, email
5.1	Jane Dixon	Office Manager	LP Appliances 215 Klondyke Dr. Outoftown, MA 21544	(W) 212-566-1234 (H) 201-567-4333 (C) 201-455-2231	jd@lpapplinc.com www.LPAppliancesonlione.com
5.2	Wendy Jensen	Director	Sunny Day School 5513 Clearview Ct. Uptown, MA 21546	(W) 212-567-1234 (H) 201-568-4333 (C) 201-456-2231	wendy.jensen@sdaysch.org www.ItsASunnyDay.com
5.3	Bob Wills	Owner	Party Hearty LLC 21 First St. Intown, MA 21544	(W) 212-432-1234 (H) 201-324-4333 (C) 201-772-2231	orders@phllc.com www.phllc.com
5.4	Jen Phillips	N/A	Home 5 Fieldview Terr. Uptown, MA 21546	(W) 212-433-1234 (H) 201-657-4333 (C) 201-545-2231	ladybug32154@home.com
5.5	Doug Braxton	N/A	Home 52 Foxrun Rd. Intown, MA 21544	(W) 212-999-1234 (H) 201-888-4333 (C) 201-770-2231	thebigbrac@jones.com
5.6	Rob Morton	N/A	Home 15 Mountainview Cir. Intown, MA 21544	(W) 212-111-1234 (H) 201-112-4333 (C) 201-113-2231	Rob21544@mmxpt.com

BCP ID	Name	Title	Company Name and Address	(W)ork Phone/ (H)ome Phone/ (C)ell Phone	Web address, email
5.7	Darnell Howe	N/A	Home 17 Mountainview Cir. Intown, MA 21544	(W) 212-727-1234 (H) 201-737-4333 (C) 201-865-2231	dlhowe@mmxpt.com
5.8	Sandy Page	N/A	Home 21 Mountainview Cir. Intown, MA 21544	(W) 212-747-1234 (H) 201-101-4333 (C) 201-947-2231	Ssp33445@mmpxt.com

6. External Contact Information (Suppliers and Business Service Providers)

A critical piece of information to have available in the event of an emergency causing interruptions to business operations is a list of suppliers and service providers to enable timely restoration of services or implementation of backup services during an interruptions to normal business operations. This list can either be compiled using the table below, or if supplier and service provider information is maintained in a separate system and a list can be printed or copied and pasted, attach it to this BCP plan as an Appendix and reference it here in place of the table.

BCP ID	Service Provided	Company	Name / Title	Account / Policy #	Phone	email
6.1	Insurance	ABC Insurance Services	Dan Thompson	AX1155-211	555-555-5555	email@email.com
6.2	Legal Services	LM Associates	Linda Miller	N/A	555-555-5555	email@email.com
6.3	Computer Services / Web Site Hosting	Company Name	First Name Last Name	Acct #	555-555-5555	email@email.com
6.4	Telephone Company (landline)	Company Name	First Name Last Name	Acct #	555-555-5555	email@email.com
6.5	Cell Phone Provider	Company Name	First Name Last Name	Acct #	555-555-5555	email@email.com
6.6	Internet Service Provider	Company Name	First Name Last Name	Acct #	555-555-5555	email@email.com
6.7	Cable / Satellite Company	Company Name	First Name Last Name	Acct #	555-555-5555	email@email.com

BCP ID	Service Provided	Company	Name / Title	Account / Policy #	Phone	email
6.8	Shipping Company	Company Name	First Name Last Name	Acct #	555-555-5555	email@email.com
6.9	Material Suppliers	Company Name	First Name Last Name	Acct #	555-555-5555	email@email.com
6.10	Travel Agency	Company Name	First Name Last Name	Acct #	555-555-5555	email@email.com
6.11	Site Maintenance Company	Company Name	First Name Last Name	Acct #	555-555-5555	email@email.com

7. Utilities and Facilities Services

A critical piece of information to have available in the event of an emergency causing interruptions to business operations is a list of utility and site maintenance companies to enable timely restoration of services or implementation of backup services during an interruptions to normal business operations, or to allow for timely notification to cut service to remove unsafe conditions at the work site (i.e. gas leak, downed power wires, etc.). This list can either be compiled using the table below, or if utility and site management information is maintained in a separate system and a list can be printed or copied and pasted, attach it to this BCP plan as an Appendix and reference it here in place of the table.

BCP ID	Service Provided	Company	Name	Title	Phone	email
7.1	Landlord / Leasing Company / Site Owner	Company Name	First Name Last Name	Acct #	555-555-5555	email@email.com
7.2	Gas Company	Company Name	First Name Last Name	Acct #	555-555-5555	email@email.com
7.3	Electric company	Company Name	First Name Last Name	Acct #	555-555-5555	email@email.com
7.4	Water Company	Company Name	First Name Last Name	Acct #	555-555-5555	email@email.com
	PO Box Rental Company	Company Name	First Name Last Name	Acct #	555-555-5555	email@email.com

8. Financial Services Contact Information

In the event of an emergency or disruption to business requiring relocation, it is imperative to maintain contact with financial institutions to enable continuation of payments to employees, suppliers and service providers, as well as processing payments from customers. This list can either be compiled using the table below, or if financial institution information is maintained in a separate system and a list can be printed or copied and pasted, attach it to this BCP plan as an Appendix and reference it here in place of the table. This information should also identify key individuals who knowledge of account numbers and have system IDs and passwords allowing access to financial data and services. It is not recommended to publish this information within the BCP plan as this may result in unauthorized access. The plan should specify a primary and secondary individual responsible for accessing financial data and services.

BCP ID	Financial Service	Contact Names (Primary and Secondary)	Address	Phone	email	Primary and Secondary Individuals with access to account, ID and password/PIN (specify Contact Tree ID from section I, 4)
8.1	Bank	Main Street Bank	50 Maint Street Downtown, MA 10255	555-555-5555	email@email.com	Susan Smith (Primary), A1.1. Steve Brown (Backup) A.1.2
8.2	Payroll Administrator	Company Name	Address	555-555-5555	email@email.com	Susan Smith (Primary), A1.1. Steve Brown (Backup) A.1.2
8.3	Benefits Administrator	Company Name	Address	555-555-5555	email@email.com	Susan Smith (Primary), A1.1. Steve Brown (Backup) A.1.2

9. Regulatory Agency Contact Information

It is critical to maintain communications with regulatory agencies during emergencies to enable reporting of business status affecting required regulatory filings or reporting.

BCP ID	Regulatory Agency	Contact Names (Primary and Secondary)	Address	Phone	email
9.1	IRS	N/A	Address	555-555-5555	email@email.com
9.2	State Tax Dept.	Joe Smith, Kate Jones	Address	555-555-5555	email@email.com
9.3	Social Security	N/A	Address	555-555-5555	email@email.com
9.4	Licensing Bureau (if applicable)	Jim Sanchez, Joel Poole	Address	555-555-5555	email@email.com

10. Critical Records and Systems

To ensure continuation of the business with minimal interruptions or negative impact to your customers, it is important to have access to critical information such as customer lists, orders, financial statements, personnel records, tax records, contracts, etc. These may reside in hard copy form or in electronic systems. For all of these records, regardless of the media or format, there must be backup copies available in the event of en emergency. Back up copies may be paper copies, microfilm, scanned files, electronic copies (CDs, DVDs, Memory stick etc.) or these may be maintained in computerized systems with remote access (i.e. Online Mailbox, online servers, etc.). The critical piece is that in the event of an emergency you can access the information you need quickly and from an alternate location. For this section, document each critical record (invoices, POs, contracts, etc.) and system (market research database, order processing system, web site server, etc.) and identify its backup. If you were to lose primary access to this information, where would you go to find it? This list should contain information needed to get to the back up record location or system (what do I need to know to be able to find this information in an emergency). It is not necessary to document system IDs and passwords here (unless you want to) but at a minimum you need to identify who has the necessary system IDs and passwords, pin, keys, combinations, etc. to be able to access the records/systems. Be sure to identify a primary and backup person here.

BCP ID	Critical Record or System	Backup Record Location / System (URL, Name&Address – how to find this information)	Primary Person With Access To This Information (Name, Contract Tree ID)	BackUp Person With Access To This Information (Name, Contract Tree ID)
10.1	Orders Database (CD)	Secure Docs Inc. 21 Main St. Downtown, MA 01234 800-555-6776	Janet Kline (A.3.3)	Stan Wilson (A.2.2)
10.2	Orders Database (online)	Online Mailbox www.OnlineMailbox.com	Janet Kline (A.3.3)	Stan Wilson (A.2.2)
10.3	Financial records / payroll	Secure Docs Inc. 21 Main St. Downtown, MA 01234 800-555-6776	Janet Kline (A.3.3)	Stan Wilson (A.2.2)

BCP ID	Critical Record or System	Backup Record Location / System (URL, Name&Address – how to find this information)	Primary Person With Access To This Information (Name, Contract Tree ID)	BackUp Person With Access To This Information (Name, Contract Tree ID)
10.4	Recipes / ingredient list	Safe Dep Box – First Bank 57 West Ave. Upper City, CT 04321 888-555-8965	Janet Kline (A.3.3)	Stan Wilson (A.2.2)
10.5	Marketing materials	Online Mailbox www.OnlineMailbox.com	Janet Kline (A.3.3)	Stan Wilson (A.2.2)
10.6	Tax Returns	Secure Docs Inc. 21 Main St. Downtown, MA 01234 800-555-6776	Janet Kline (A.3.3)	Stan Wilson (A.2.2)
10.7	Real Estate documents	Secure Docs Inc. 21 Main St. Downtown, MA 01234 800-555-6776	Janet Kline (A.3.3)	Stan Wilson (A.2.2)
10.8	Contracts	Secure Docs Inc. 21 Main St. Downtown, MA 01234 800-555-6776	Janet Kline (A.3.3)	Stan Wilson (A.2.2)
10.9	Website files	Secure Docs Inc. 21 Main St. Downtown, MA 01234 800-555-6776	Janet Kline (A.3.3)	Stan Wilson (A.2.2)

11. Backup Locations

To ensure continuation of critical business functions during activation of the BCP plan, backup/contingency locations must be identified for all essential operations which cannot be performed remotely (at home). Examples are laboratory or manufacturing operations, and conference facilities for meeting with customers. Backup sites should be far enough away from the primary work site to reduce the chances that the disaster will affect access to the backup site, but not so far as to prevent timely access by personnel. A general rule is for the backup site to be 20-50 miles from the primary site. Backup site information in the BCP plan must include directions instructing personnel how to get to the location. This list can either be compiled using the table below, or if backup location information is maintained in a separate system and a list can be printed or copied and pasted, attach it to this BCP plan as an Appendix and reference it here in place of the table.

BCP ID	Function / Operation	Backup Location Name*	Contact Names (Primary and Secondary)	Address	Phone	email
11.1	Food storage	GB Bakery	Kyle Jones (Primary) William Farley (Secondary)	512 Spring St. Leftville, CT 05562	547-234-5555 (W) 547-555-4333 (C)	kjones@gbb.com wfarly@gbb.com
11.2	Kitchen / food preparation / baking	ACE Kitchens	Tom Johnston (Primary) Jill Gleeson (Secondary)	2112 Vine St. Leftville, CT 05562	567-432-7867 (W) 567-555-9353 (C)	tom.johnston@abl.com jill.gleeson@abl.com
11.3	Delivery	S. Smith - Home	Susan Smith (Primary) Dan Smith (Secondary)	12 Mountaintop Dr. Sunnyville, CT 05573	201-555-4333 (H) 201-444-2231 (C)	ssmith@ekcllc.com dsmith@xbc.com
11.4	Customer Service / order processing	S. Smith - Home	Susan Smith (Primary) Dan Smith (Secondary)	12 Mountaintop Dr. Sunnyville, CT 05573	201-555-4333 (H) 201-444-2231 (C)	ssmith@ekcllc.com dsmith@xbc.com

* **Directions** and **Maps** to all backup locations are included in the Appendix Section of this BCP.

12. Backup Service Provider and Supplier Information

In the event of an emergency or disruption to primary business services and suppliers, a list of back up services and suppliers must be maintained for quick reference to enable restoration of essential business functions in a timely manner. This list can either be compiled using the table below, or if backup services and supplier information is maintained in a separate system and a list can be printed or copied and pasted, attach it to this BCP plan as an Appendix and reference it here in place of the table.

BCP ID	Service / Supplier	Company	Contact Names (Primary and Secondary)	Address	Phone	email
12.1	Ingredients and food containers	AOK Food Service Supplies	Tony Price (Primary) Dale Watson (Secondary)	2931 Spring St. Leftville, CT 05562	547-234-5555 (W) 547-555-4333 (C)	orders@aok.com
12.2	Computer services / web hosting	AQXBZ.com	David Fine (Primary) Amit Patel (Secondary)	17 Vine St. Leftville, CT 05562	567-432-7867 (W)	web@aqxbz.com
12.3	Payroll administration	ChecksInTheMail Inc.	Donna Phillips (Primary) Steve Harris (Secondary)	2912 Mine St. Sunnyville, CT 05573	547-234-5555 (W) 547-555-4333 (C)	donna@citm.com steve@citm.com
12.4	Fresh fish delivery	Fishermart	Clive Barron (Primary) Tom Cox (Secondary)	98 Mont Blanc Dr. Sunnyville, CT 05573	567-432-7867 (W) 567-555-9353 (C)	N/A
12.5	Cleaning	Doris Bloom & Co. LLC	Doris Bloom (Primary)	12 Millstone Terr. Sunnyville, CT 05573	567-432-7867 (W) 567-555-9353 (C)	N/A

13. Emergency Services Contact Information

In any emergency situation, it is essential to know who to contact. This information must be on hand for easy availability. List all critical emergency contact information in the table below so you have it all in once place for easy access.

BCP ID	Emergency Service	Contact Name(s) (if it is important)	Address (if it is important)	Phone (include direct numbers if known, in case 911 services are unavailable)	Email / Web
13.1	Local Police Dept.	N/A	15 Main St. Downtown, MA 01234	911 or 1-855-555-4345	police@downtown.ma.us
13.2	State Police	N/A	N/A	911 or 1-800-222-6666	N/A
13.3	Fire Dept.	N/A	N/A	911	N/A
13.4	Local Emergency Management Director	Capt. Tom Harding	15 Main St. Downtown, MA 01234	911 or 1-855-555-4345	tharding@downtown.ma.us
13.5	First Aid Squad	N/A	N/A	911	N/A
13.6	Hazmat	N/A	N/A	911	N/A
13.7	FBI	N/A	N/A	1-800-225-5324	http://www.fbi.gov/
13.8	Homeland Security	N/A	N/A	1-202-282-8000	http://www.dhs.gov/
13.9	Poison Control	N/A	N/A	1-800-222-1222	http://www.aapcc.org/
13.10	Local Township/Municipality	N/A	15 Main St. Downtown, MA 01234	855-555-6784	downtown@downtown.ma.us

Section II - Business Risk Assessment and Impact Evaluation

This section of the BCP plan documents the risk associated with the disruption of any key business operation or process. Risk is defined as the combination of how likely a key operation will be disrupted, how much time before the business experiences the negative impact of losing the operation, and how much this disruption will hurt the business performance. (*Ex. The manufacturing plant can be shutdown for several months in the event that the river running through the property overflows and floods the building. Some flooding has occurred once every 3 years with major flooding occurring on average every 25 years. This will result in loss of sales within 2 days*).

The information documented in this section is used to justify the prioritization of the BCP plan activities and justify the money allocated to ensuring the continued operations during an emergency or disaster. (*Ex. A backup manufacturing site is qualified and PO-##### is issued to Company Z to enable start up of production within 1 day in the event of an emergency at the primary manufacturing facility. A new manufacturing site has been identified and a project initiated to relocate primary manufacturing to this site by 15-JUN-20##*).

The more these figure can be quantified, the better will be the BCP decision-making.
To assist in this evaluation, use the scale indicated in the chart below.

Once this section is complete, this information will provide a roadmap for prioritizing the BCP tasks.
The result will be a list of key business operations/processes with risk levels assigned. The operations/processes with the lowest risk numbers will have the highest risk level. The high risk items are the ones which should receive the most attention when designing and implementing the BCP plan. These are the items which will hurt the business the most if they are not adequately addressed.

14A. Business Process Risk Evaluation

The critical elements of completing this section are:

- Identifying the key business operations and processes. These are operations or process which must be working in order for the business to continue.
- Identifying the likelihood of disruption (chances that something unexpected could happen which disrupts the key business operation or process). **SCALE: 1 – 10** (1 = Very High Likelihood of Unplanned Disruption, 10 = Very Low Likelihood of Unplanned Disruption)
- Identifying how significant the effect is of losing this operation or process. (i.e. people at risk, profits lost, customers lost) **SCALE: 1 – 10** (1 = Very High Significance To The Business, 10 = Very Low Significance To The Business). Identifying how long this operation can be disrupted before it negatively impacts the business. **Number of Days** without this operation before business would experience negative impact to operations and/or sales.

BCP ID	**Key Business Process (Operation/Process which must be working for the business to continue)**	**(A) Likelihood Of Disruption SCALE: 1 (High) - 10 (Low)**	**(B) Significance Of Negative effect if disrupted for more days than identified in (C) SCALE: 1 (High) - 10 (Low)**	**(C) Time Before Negative Business Effect (Number of Days)**	**BCP Priority = (A) x (B) + (C) (Lower Number = Higher Risk and Higher BCP Priority)**
14.1	Kitchen operations	1	3	3	6
14.2	Food Storage	3	3	0	9
14.3	Delivery/distribution	5	4	2	22
14.4	Customer service	8	1	0	8

14B. Business Process Disruption Impact Evaluation

For the key business processes identified in section 14A, evaluate the impact to the business of an unplanned disruption in terms of:

- Available work-arounds to temporarily replace the process *(ex. paper records can be processed manually in the event that a key computer system is unavailable, process can be outsourced to Company X).*
- How long can the business continue to operate until the process must be restored? *(ex. computer system must be back online within 3 months or costs and loss of efficiency will reach unsustainable levels).*
- What resources are needed for work-around and for restoration of the process? *(ex. for the duration of manual processing of paper forms, resources needed include: 3 temporary people working 8 hours per day each, 3 computers with internet access, 1 printer, XYZ software with 3 licenses, workspace for 3 people, 1 telephone, 1 FAX, etc.).*

List the Key Business Processes in order of BCP Priority (lowest number first) using the information identified in the table in question 14A. The objective here is to identify which processes require the most immediate attention in the event of a disruption and what is required to address the disruption.

BCP ID (From section 14A.)	Key Business Process	Work-arounds available to temporarily replace this process.	How long can business operate until process must be restored? (Number of Days)	Resources needed to temporarily replace this process. (People, Services, Equipment, Facilities, Money, etc.)	Resources needed to restore this process. (People, Services, Equipment, Facilities, money, etc.)
14.1	Kitchen operations	Outsource to ACE Kitchens	5	- Manager to oversee - Check for $500 per day - 2 delivery vehicles	- Manager to oversee - Kitchen staff (3 people) - 2 delivery vehicles

BCP ID (From section 14A.)	Key Business Process	Work-arounds available to temporarily replace this process.	How long can business operate until process must be restored? (Number of Days)	Resources needed to temporarily replace this process. (People, Services, Equipment, Facilities, Money, etc.)	Resources needed to restore this process. (People, Services, Equipment, Facilities, money, etc.)
14.2	Customer service	Operate remotely	0	- 2 people - Internet connection - (2) laptops with IE8 - Cell phone - Access to voice mail	- Office staff (2 people) - Office with working utilities and internet connection - 2 computers with IE8 - phone system linked to 800 number
14.3	Food Storage	Outsource to GB Bakery	1	- Manager to oversee - Check for $350 per day - 1 delivery vehicle	- Kitchen staff (1 person) - Facility with security and working utilities - Clean area - working refrigerator and freezer - Pantry area with shelves (200 sq.ft.)

BCP ID (From section 14A.)	Key Business Process	Work-arounds available to temporarily replace this process.	How long can business operate until process must be restored? (Number of Days)	Resources needed to temporarily replace this process. (People, Services, Equipment, Facilities, Money, etc.)	Resources needed to restore this process. (People, Services, Equipment, Facilities, money, etc.)
14.4	Delivery/distribution	Use personal vehicles from home	5	- At least 4 SUV/Van size vehicles located within 10 miles of downtown. - Fuel - 4 GPS and local maps - 4 cell phones - Customer list with complete address and phone - 4 carts with hand trucks	- Functional warehouse area with secure storage - Food storage capabilities - Functional delivery vehicles - Roadways open

Section III - Pre-Emergency Planning

15. Risk Mitigation

Identify any risk mitigation options are available to reduce the chances of an unplanned disruption occurring and/or reduce the negative impact to the business. *(Ex. Create 2 backup copies of sales forecast and customer database and store in 2 different secure offsite locations. Location 1 is __________, location 2 is _________).*

BCP ID	Key Business Process	Risk Mitigation Plan (if none possible, indicate "none")
15.1	Kitchen operations	Set up agreement with ACE Kitchens
15.2	Customer service	Obtain and configure laptops and cell phones for use at home. Route 800 calls to roll over to cell phone if primary phone not answered.
15.3	Food Storage	Obtain generator capable of running 1 freezer.
15.4	Delivery/distribution	None available.

16. Emergency Materials and Supplies

For each Key Business Process, identify all critical emergency materials and supplies that need to be available in the event of an emergency or unplanned business interruption. This includes small disposable supplies (i.e. pens, paper, mailing supplies, etc.), office equipment (computers, printers, phones, etc.), production equipment (i.e. machinery needed to service customers, refrigerator, freezer, generator, batteries, tools, etc.) as well as raw materials and production supplies (i.e. paper, paint, lawn fertilizer, baking ingredients, gasoline, propane, etc.).

BCP ID	Key Business Process	Materials / Supplies needed (Location or Contact information)
16.1	Kitchen operations	- 2 Carts with hand trucks (warehouse) - At least 2 SUV/Van sized vehicles (S. Smith and D. Jones home) - 50 Clean empty cartons (warehouse) - Packaging tape, permanent marker, blank labels (main office) - Service agreement with ACE Kitchens (office safe – S. Smith, D. Jones have combination) - 5 large coolers (warehouse) - Towels, hand sanitizer (break room storage closet)
16.2	Customer service	- 2 Spare laptops (S. Smith home) - UPS power backup (S. Smith home) - 1 Internet card (S. Smith home) - 2 Cell phone chargers (S. Smith home)
16.3	Food Storage	- 5,000 Watt generator (warehouse) - Keys to GB Bakery back door (D. Jones, K. Greene)
16.4	Delivery/distribution	- 4 (10)-gallon Fuel containers (warehouse, S. Smith and D. Jones home) - 2 Carts with hand trucks (warehouse) - 2 local maps and GPS (S. Smith and D. Jones home) - Customer list (online at Online Mailbox - S. Smith, D. Jones have account information)

Section IV - BCP Plan Activation and Implementation

In this section, identify the triggers for activating the BCP plan (what events must occur in order to activate the BCP plan), who is involved in activating the BCP plan, and what are the steps to activate and implement the plan. All key business processes identified in Section 14A should be listed in Section 16 in order of BCP priority.

Start times should be specified based on impact evaluation information captured in section 14B (*ex. if production line downtime causes loss of customer orders within 2 days, then trigger is that production line is down for more than 1 day and target start time for production operation BCP should be specified to have a backup process in place within 2 days*).

17. Plan Activation

BCP ID	Key Business Process	BCP Activation Trigger	Responsible Individual(s) for Activating / Implementing BCP	BCP Plan ID (refer to section 18)	BCP Contact Tree* (refer to sections 3 and 4)
17.1	Kitchen operations	- Kitchen inoperable for more than 2 workdays. And - 2 or more orders pending.	- S. Smith (primary) - S. Brown (secondary)	18.1	A, B
17.2	Customer service	- Communications with customers unavailable for more than 8 consecutive hours.	- S. Smith (primary) - S. Brown (secondary)	18.2	A, C, D

BCP ID	Key Business Process	BCP Activation Trigger	Responsible Individual(s) for Activating / Implementing BCP	BCP Plan ID (refer to section 18)	BCP Contact Tree* (refer to sections 3 and 4)
17.3	Food Storage	- Refrigeration/Freezer systems off line for more than 4 hours. Or - Access to primary food storage area is not available for more than 1 business day.	- S. Smith (primary) - S. Brown (secondary)	18.3 - Plan A (No power to refrigeration systems) 18.4 - Plan B (No access to refrigeration systems / refrigeration not operational)	A, B
17.4	Delivery/distribution	- Delivery service is unavailable for more than 4 days. Or - Roads are blocked for more than 4 days.	- S. Smith (primary) - S. Brown (secondary)	18.5	A, D

*Upon activation of the BCP plan initiate the contact tree (Refer to Sections 3 and 4) to account for all personnel and to alert team members of BCP activation

18. Plan Implementation

In this section, define the tasks needed to start operating under the BCP plan. For each task, document who is assigned the responsibility for completing the task. The tables below are set up to document the tasks, who is responsible and a section for verification that the task was completed. As an alternative to these tables, the implementation plan can be a separate document attached to this document as an Appendix.

18.1 Key Business Process: Kitchen operations

Start Time from BCP Activation (Days, Hours)	BCP Task	Assigned To	Completion Verification (enter Initials / Date / Time upon completion)
0 Hours	Activate Contact Trees A, B and confirm contact.	Jill Jenkins, Linda Hudson	
2 Hours	Contact ACE Kitchens and confirm availability and payment schedule.	Jill Jenkins, Linda Hudson	_______ / _______ / _______
2 Hours	Confirm Food Storage and Customer Service BCP plans are activated.	Jill Jenkins, Linda Hudson	
3 Hours	Review pending orders with staff and with Ace. Create action plan and schedule for obtaining ingredients and transporting finished orders to customers.	Jill Jenkins, Linda Hudson	
6 Hours	Provide customers with delivery schedule.	Jill Jenkins, Linda Hudson	
8 Hours	Provide ACE with prioritized list of orders, recipes, and necessary ingredients if not available at ACE location.	Jill Jenkins, Linda Hudson	

Start Time from BCP Activation (Days, Hours)	BCP Task	Assigned To	Completion Verification (enter Initials / Date / Time upon completion)
12 Hours - Ongoing	Coordinate food preparation orders at ACE. Oversee shipments of ingredients and finished orders to and from ACE.	Jill Jenkins, Linda Hudson	

18.2 Key Business Process: Customer service

Start Time from BCP Activation (Days, Hours)	BCP Task	Assigned To	Completion Verification (enter Initials / Date / Time upon completion)
0 Hours	Activate Call Tree A, C, D and confirm	Sam Lockwood, Sue Cross	
1 Hour	- Set up laptops. - Open web enabled email service. - Create online customer call log to track incoming phone calls	Sam Lockwood, Sue Cross	_______ / ________/ ________
2 Hours - Ongoing	- Field incoming calls, online orders and emails. - Communicate with customers as needed the status of orders. - Inform suppliers of any changes due to BCP activation.	Sam Lockwood, Sue Cross	

18.3 Key Business Process: Food Storage – Plan A (No power to refrigeration systems)

Start Time from BCP Activation (Days, Hours)	**BCP Task**	**Assigned To**	**Completion Verification (enter Initials / Date / Time upon completion)**
0 Hours	Activate Contact Trees A, B and confirm contact.	Jill Jenkins, Linda Hudson	
2 Hours	Start up generator and confirm refrigeration is operational.	Jill Jenkins, Linda Hudson	_______ / ________ / ________
3 Hours - Ongoing	- Monitor refrigeration temperature and fuel level. - Obtain fuel as needed to supply generator. Generator must be refueled when tank level reaches ¼ or less.	Jill Jenkins, Linda Hudson	
8 Hours - Ongoing	Activate BCP 18.4 in the event that refrigeration fails for any reason.	Jill Jenkins, Linda Hudson	

18.4 Key Business Process: Food Storage – Plan B (No access to refrigeration systems / refrigeration not operational)

Start Time from BCP Activation (Days, Hours)	BCP Task	Assigned To	Completion Verification (enter Initials / Date / Time upon completion)
0 Hours	Activate Contact Trees A, B and confirm contact.	Kate Greene, Janet Kline	
2 Hours	Contact GB Bakery and confirm availability and payment schedule.	Kate Greene, Janet Kline	_______ / ________/ ________
4 Hours	Arrange shipment of ingredients and finished orders to and from GB.	Kate Greene, Janet Kline	
8 Hours - Ongoing	- Deliver ingredients and finished orders to GB. - Coordinate with Kitchen on delivering ingredients to ACE or internal kitchen. - Coordinate with Customer Service on shipping orders to customers.	Kate Greene, Janet Kline	

18.5 Key Business Process: Delivery/distribution

Start Time from BCP Activation (Days, Hours)	BCP Task	Assigned To	Completion Verification (enter Initials / Date / Time upon completion)
0 Hours	Activate Call Tree A, D and confirm	Kate Greene, Janet Kline	_______ / ________/ ________
2 Hours	- Obtain 2 or more carts with hand trucks. - Arrive at S. Smith home with 2 or more SUV/Van sized vehicles.	Kate Greene, Janet Kline	
3 Hours	Coordinate with customer service on shipping/delivery schedule for ingredients and finished orders.	Kate Greene, Janet Kline	
4 Hours	Assign routes to drivers and vehicles.	Kate Greene, Janet Kline	
8 Hours - Ongoing	- Deliver ingredients to kitchen or ACE. - Deliver finished order to customers. - Monitor fuel level and obtain fuel as needed.	Kate Greene, Janet Kline	

Section V - BCP Plan Deactivation and Restoration of Normal Operations

Once the emergency situation has passed or is no longer an impact to the business, it is important to know what to do to get back to normal business operations and when to start doing this. This information needs to be defined before an emergency happens.

19. BCP Plan Deactivation

In this section, document the BCP De-Activation Trigger for each Key Business Process; what would need to happen in order to stop following the BCP plan and go back to conducting business like you did before the emergency (Ex. *Utilities at primary work location have been restored*). Be sure to document who is responsible for deactivating the BCP and initiating the restoration plan. Also be sure to document the affected Contact Tree (i.e. who needs to be informed of the deactivation of the BCP and initiation of the restoration).

BCP ID	Key Business Process	BCP De-Activation Trigger	Responsible Individual(s) for De-Activating BCP and initiating Restoration Plan	Restoration Plan ID (refer to section 20)	BCP Contact Tree* (refer to sections 3 and 4)
19.1	Kitchen operations	All of these are met - - Main kitchen is accessible - Kitchen has power, heat, A/C, running water. - Kitchen is clean and safe to work in. - Supplies and equipment are available - Kitchen staff is present (at least 2 people)	- S. Smith (primary) - S. Brown (secondary)	20.1	A, B

BCP ID	Key Business Process	BCP De-Activation Trigger	Responsible Individual(s) for De-Activating BCP and initiating Restoration Plan	Restoration Plan ID (refer to section 20)	BCP Contact Tree* (refer to sections 3 and 4)
19.2	Customer service	All of these are met - - Main office is accessible - Main office has power, heat, A/C, running water. - Office is safe to work in. - Computers are functional and online. - Phones are working. - Printers are working. - Personnel (at least 2) can get to office.	- S. Smith (primary) - S. Brown (secondary)	20.2	A, C, D
19.3	Food Storage	All of these are met - - Refrigerators and freezers operational (refrigerators at 38-45 deg, freezer below -5 deg) - Area is safe to work in. - Area is accessible.	- S. Smith (primary) - S. Brown (secondary)	20.3	A, B

BCP ID	Key Business Process	BCP De-Activation Trigger	Responsible Individual(s) for De-Activating BCP and initiating Restoration Plan	Restoration Plan ID (refer to section 20)	BCP Contact Tree* (refer to sections 3 and 4)
19.4	Delivery/distribution	All of these are met - - Roadways to warehouse are open. - Warehouse is operational and has power, heat, A/C, running water. - Warehouse is safe to work in. - Carts and hand trucks are available. - Vehicles are operational and fuel is available. - At least 2 drivers are available.	- S. Smith (primary) - S. Brown (secondary)	20.4	A, D

20. Return To Normal Operations

In this section, define the tasks needed to move from operating under the BCP plan to restoring normal operations. For each task, document who is assigned the responsibility for completing the task. The tables below are set up to document the tasks, who is responsible and a section for verification that the task was completed. As an alternative to these tables, the restoration plan can be a separate document attached to this document as an Appendix.

20.1 Key Business Process: Kitchen operations

Restoration Task	**Assigned To**	**Completion Verification (enter Initials / Date / Time upon completion)**
Inspect kitchen for safety.	Jill Jenkins, Linda Hudson	________ / ________ / ________
Verify utilities (power, heat, A/C, running water) are functional.	Jill Jenkins, Linda Hudson	
Check equipment is available and functional.	Jill Jenkins, Linda Hudson	
Bring in staff (at least 2 people).	Jill Jenkins, Linda Hudson	
Clean kitchen.	Jill Jenkins, Linda Hudson	
Arrange to move ingredients to kitchen.	Jill Jenkins, Linda Hudson	
Verify supplies against pending orders and purchase new ingredients and supplies if needed.	Jill Jenkins, Linda Hudson	

20.2 Key Business Process: Customer service

Restoration Task	Assigned To	Completion Verification (enter Initials / Date / Time upon completion)
Verify office is open and accessible.	Sam Lockwood, Sue Cross	________ / _________ / _________
Verify utilities (power, heat, A/C, running water) are functional.	Sam Lockwood, Sue Cross	
Verify computers and phone are operational.	Sam Lockwood, Sue Cross	
Test Internet connection.	Sam Lockwood, Sue Cross	
Test printer.	Sam Lockwood, Sue Cross	
Verify levels of office supplies – order as needed.	Sam Lockwood, Sue Cross	
Bring in staff (at least 2 people).	Sam Lockwood, Sue Cross	

20.3 Key Business Process: Food Storage

Restoration Task	Assigned To	Completion Verification (enter Initials / Date / Time upon completion)
Inspect area for safety.	Jill Jenkins, Linda Hudson	________ / ________ / ________
Verify utilities (power, heat, A/C, running water) are functional.	Jill Jenkins, Linda Hudson	
Check operation of freezer (must be below -5 deg) and refrigerator (must be 38-45 deg).	Jill Jenkins, Linda Hudson	
Bring in staff (at least 2 people).	Jill Jenkins, Linda Hudson	
Clean freezer and refrigerator – exterior and interior.	Jill Jenkins, Linda Hudson, Phil Smith	
Arrange to move ingredients and finished orders to food storage area.	Jill Jenkins, Linda Hudson, Phil Smith	

20.4 Key Business Process: Delivery/distribution

Restoration Task	**Assigned To**	**Completion Verification (enter Initials / Date / Time upon completion)**
Inspect warehouse area for safety.	Kate Greene, Janet Kline	________ / ________ / ________
Verify availability of all delivery vehicles and drivers.	Kate Greene, Janet Kline	
Obtain customer order list.	Kate Greene, Janet Kline	
Create delivery schedule, coordinated with Food Storage and Kitchen BCPs / Restoration Plans.	Kate Greene, Janet Kline	

Section VI - BCP Plan Testing

To ensure that a BCP plan will be effective in the event of an emergency, it is important to test it out. Just like a fire drill tests out how good your plan is for evacuating the area in an emergency and how well people are trained and prepared to respond to an evacuation, you need to know how well your BCP plan is set up and how well people can follow it. Testing will identify any areas that need attention or revising. Ideally, the test should be documented in a report which is attached to the BCP plan to document what was done, what the results were and any follow up actions that need to be taken to ensure effectivity of the BCP plan.

BCP Plan Test Date	Completed By (Printed Name/Signature/Date)
10-AUG-2011	S. Smith / S. Smith / August, 10, 2011

Section VII - BCP Plan Maintenance

It is important to keep a BCP plan current and up to date in order for it to be useful when you really need it. This means that the plan must be reviewed and information updated on a regular or as-needed basis. This can either be done every year or every 6 months as it makes sense, or it can be every time there is a change that impacts on the information contained in the BCP plan.

When the plan is reviewed, indicate the date the plan was reviewed and what changes, if any, were made. If the plan is reviewed on a regular time basis, then it is possible that the review will find that all of the information is current and correct and thus no changes are necessary. It is still important to indicate the date and completion of this review to specify that the plan is still current.

BCP Plan Review Date	Updates / Changes Made, if any. (brief description of sections updated)	Completed By (Printed Name/Signature/Date)
12-AUG-2011	Revised Restoration plan for Delivery to coordinate schedule with Food Storage and Kitchen BCP / Restoration plan.	Steve Brown / Steve Brown / August, 12, 2011

Section VIII - Appendices / Attachments

Appendix I - Map and Driving Directions to ACE Kitchens Inc.

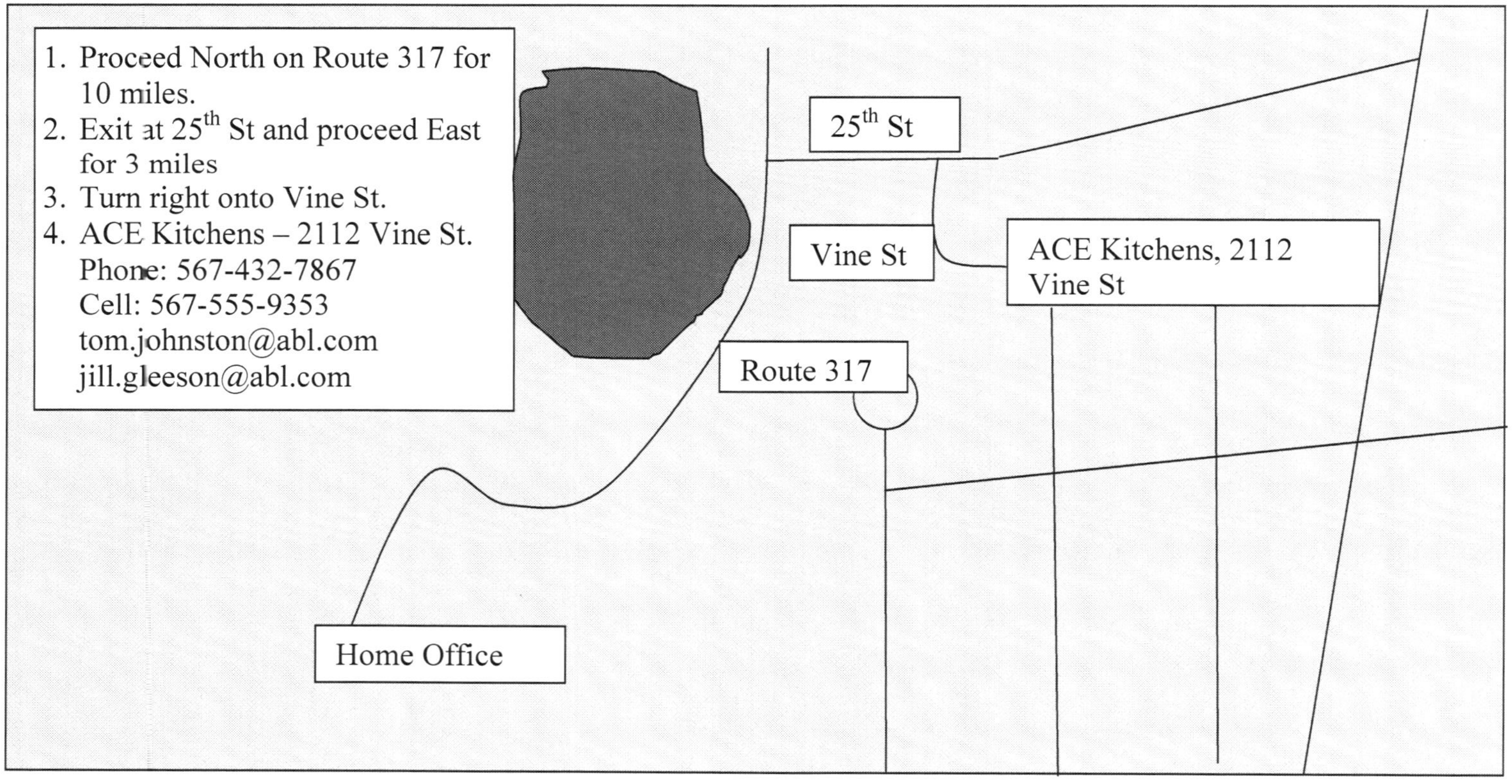

Appendix II - Instructions for Loading and Unloading the Freezer at GB Bakery.

1. Obtain key to Freezer #5 from shift supervisor upon entering GB Bakery.

   Shift supervisor station phone number is 612-555-9984.
2. Bring in carts from rear bay door #3 (on the left of the parking lot).
3. Load shelves in the rear of Freezer #5 from the bottom up, starting with the heaviest items.
4. Check the temperature (look at the read-out). And adjust if necessary to -7 Deg F.
5. Fill in the log sheet hanging from the freezer door exterior to document all EK Catering items being stored in the freezer.
6. As items are being removed, note this in the log sheet. Include item, quantity, name, time, and date.

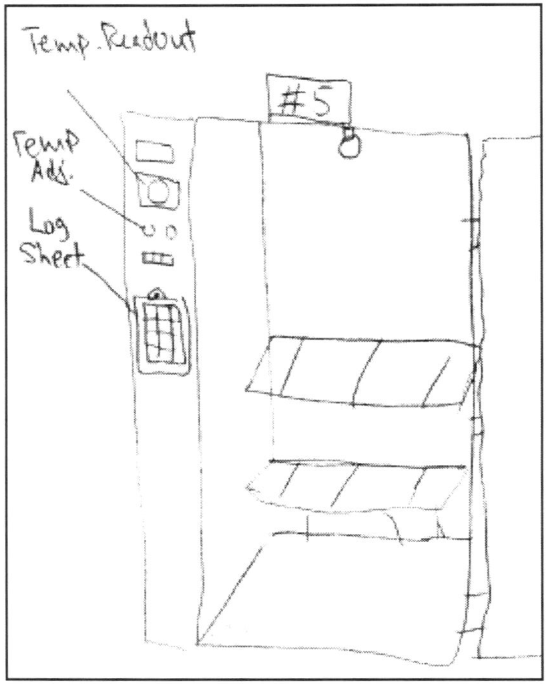

BCP (Business Continuity Plan) Essentials

Table of Contents

The Importance Of Continuity Planning For Any Business (Even Small Businesses) 1
Critical Elements of a Business Continuity Plan (What Every Successful BCP Must Have) 3
Top Reasons Why BCPs Fail and How To Make Sure Yours Does Not 4
How To Get Started Setting Up An Effective BCP 7

The Importance Of Continuity Planning For Any Business (Even Small Businesses)

A Business Continuity Plan (BCP) is a roadmap to enable a business to continue operations under adverse conditions, such as an unforeseen disaster or other unplanned interruption to the business. This includes:

- Natural Disasters, such as: bad weather, flooding, earthquake, disease, etc.
- Man-made business interruptions, such as: labor strikes, power outages, arson, theft, terrorism, trade embargo, etc.
- Any other identified risk to the business, such as: critical supplier closes their doors unrepentantly

Business continuity planning is a process of identifying the potential risks to your business and then evaluating how to prepare for these so that if they do happen, your business can continue to be viable. This means that you are still able to operate at some level needed to meet customer needs and to be able to resume normal operations at a defined point in the future. Lack of adequate BCP means that a disaster could put you out of business permanently.

This is not the same thing as IT disaster recovery. IT disaster recovery involves making sure your IT systems can continue to function and your data is still available in the event that your primary systems go down; if a fire wipes out your server room for example or a massive blackout disables your primary network routers. Typically your IT service provider will provide you with options for backup hosting at other sites as part of Disaster Recovery Planning (DRP). Large companies with big IT departments in different locations can manage DRP internally by having backup servers at different sites.

DRP is an important part of BCP, but BCP goes beyond just IT systems and looks at the entire business structure. It is critical that the IT system continue to function after a disaster, but if there is no one available or able to use these systems then how will the business continue to operate? BCP seeks to answer the big-picture questions such as:

- In the event of a disaster where will your people work from?

- If downed trees and power lines cut off major roadways needed for distributing your product what will you do?
- If a swine flu outbreak keeps 3/4 of your staff at home can your business still function?

This is what BCP looks at; being prepared for the unplanned interruptions. Making advanced preparations just in case of disaster is good business, but it does involve some cost, which can be significant. This is where cost-conscious organizations and businesses with limited budgets have major concerns. The result can be avoidance or undue risk-taking; hoping for the best because it just costs too much to be adequately prepared. This need not be the case. Spending a lot of money to set up a BCP does not make it more value added nor better for your business. Taking a logical approach that makes sense for your business and understanding the risks does make a BCP more value-added.

To be effective and value-added the level of preparation needs to be evaluated against the potential risk. This is what a good BCP plan will evaluate. The plan needs to be created after carefully looking at each critical business operation and identifying the associated level of risk should this operation go down. In addition, there needs to be a thorough evaluation of the options for maintaining business operations in the event of a disaster; i.e. backup or contingency plans. Based on this evaluation (typically know as a Business Impact Analysis or BIA), an effective and realistic BCP can be created which will effectively help the business to survive and operate during an unforeseen interruption or disaster.

<u>The critical elements of the plan must include</u> not just the IT systems but also:

- Personnel/employees: where will people work from? How will they do their work?
- Production: Where will you produce your products if your main facility is down? Is there a qualified backup facility? How quickly can it come online? Do you need to have inventory on hand? Do you need a contract with this facility, just in case you need them?
- Distribution: How will you get your product to your customers if your facility is down? Or if roads are inaccessible?
- Communications (both internal and external) with customers, vendors, media, local authorities and governmental regulators – **this is one of the most critical elements of an effective BCP**.

All of these details need to be considered up front before a disaster happens. Think about all your daily recordkeeping and administrative tasks. If you do not pay your taxes or your bank loans because your records are under water, you may not be in business after the water level goes down. If you cannot pay suppliers, you may not have products to make. If you cannot communicate with customers, they may go to competitors.

Big businesses have readily adopted this strategy and brought in consulting firms and purchased enterprise software to set up and manage BCPs. This has created the biggest opportunity for disaster-planning consultants since Y2K. The down side is that the price tag for having a firm set up and manage a BCP can go into the $100,000's. For small businesses, this is too cost prohibitive but BCP is still critical and this should not deter small businesses from seriously thinking about what they would do in the event of a disaster.

Rather than paying out big$$$ for a team of consultants or licensing enterprise software, small businesses can benefit from following the same logical thinking that the consultants use.

BCP firms typically come into a client with a standard set of templates (a basic generic roadmap for the BCP) which identifies all the potential risks and key operations in a business (paying vendors, manufacturing, customer service, regulatory filings, etc). The consultants then interview key individuals in each area and ask them how important each operation they perform is (how long can the business survive with out it?). Then they ask these same people what the options are to keep this operation going if a disaster happens (can you outsource it? To whom? Can you buy extra supplies and keep them somewhere else?).

Then with this information, the consultants modify the standard templates as needed to suit the situation. The final modified templates are presented as a customized BCP plan to the business management, who ultimately makes the decisions on what they will support in terms of what steps will be put into place and how much money will be spent.

The bulk of the specific actionable information in the BCP comes from the business, not the consultants. The BCP templates provide a guide to prompt the appropriate discussions and ask the right questions. So armed with a basic BCP template, a small business owner can perform this same procedure and come up with an effective BCP on their own or with the help of conscientious person (not necessarily a BCP consultant).

Although the process of setting up and managing a BCP seems very complicated and/or prohibitively expensive for small businesses, it does not need to be. There are materials and resources available that would greatly help out the small businesses to be prepared for the unforeseen without the need to spend large sums of money.

Critical Elements of a Business Continuity Plan (What Every Successful BCP Must Have)

So what are the standards for Business Continuity Planning? If you boil down any of the BCPs and templates floating around out there, you will find there are certain critical elements which all must address. If you can address these critical elements in your plan, then it will be effective.

The critical elements of a BCP which need to be addressed include:

(1) Data back-up and recovery (critical information such as customer records, tax records, etc. Both in hard copy and electronic).
(2) Mission critical systems (IT systems needed to conduct business).
(3) Internal communications with employees.
(4) External communications with emergency services, media, service providers, regulatory agencies and vendors.
(5) Customer service.
(6) Facilities needed for production, manufacturing, warehousing and distribution.
(7) Physical work locations of employees.
(8) Identification of BCP roles and responsibilities.
(9) Objectives for activation, deactivation and recovery times for critical business operations.
(10) Ongoing testing and review of the BCP.

These are the big ticket items that you must have answers to before a disaster happens. If you can get as handle on these, your BCP will provide you with value added information to help get you through a disaster.

Top Reasons Why BCPs Fail and How To Make Sure Yours Does Not

A Business Continuity Plan (BCP) is a valuable tool for being prepared in the event of an unforeseen disaster which impacts your business. Most organizations will readily agree with this fact that a BCP is needed. However there is great disagreement within organizations when the question arises "how much planning is enough?". A BCP is a great hedge against unforeseen business interruptions, but it does come at a cost. Creating, revising and testing a BCP costs both time and money; time of employees spent writing and testing the plan, plus money spent on contracts with backup vendors and consultants. This is when management begins to question the value of BCP and decisions need to be made to balance the scope of contingency planning against the cost. At what point are you spending too much for the value you are receiving?

Part of this question is answered in the risk assessment and business impact analyses phase of the BCP creation. If created appropriately, items identified to be of sufficient risk to the business need require contingency planning. The scope (and cost) of contingency planning should be proportional to the risk. For example, an informational website may not require the same level of redundancy and recovery time objectives as an order-tracking system. This is evaluated by the business and supported by results of testing the BCP.

But beyond this, how to you know if the BCP is appropriate? Is there more you can do...without doing too much? An excellent way to gain insight into the robustness of your particular BCP is to look at the most common types of BCP failures; situations where the BCP did not ensure the business recovery time objectives. Take a look at these common reasons why BCPs fail to meet objectives and then carefully evaluate your BCP against these areas.

So what are the most common reasons why BCPs do not live up to expectations?

1. The reasons for implementing the BCP are not supported at the top.

I have been involved in organizations where BCPs were being implemented because either it sounded like the right thing to do or someone thought it was a good idea because the auditors were rescheduled to show up soon. The problem was that senior management wasn't really on board with this idea and therefore no one beneath them really believed this was worth spending the time or money on. The result of this is a great looking set of documentation with very little behind it. Example: BCP says "if the site is inaccessible, all employees work from home and we go to a contact manufacturing site." This sounds great but is this feasible? If there is no IT infrastructure in place to enable everyone to work from home, and management in not willing to negotiate a contract for contingency services with contract manufacturers, because both of these items have a cost, then the plan is doomed to fail if you really needed it.

2. BCP does not look at the entire business.

Many organizations confuse Disaster Recovery Planning (DRP) with Business Continuity Planning (BCP). DRP focuses on IT systems while BCP focuses on the entire business and organization. It is great to be able to recover the IT systems, but if people are not available to use them nor manufacture products nor deliver them to customers, then the plan is not complete.

3. Owner of the BCP is not clearly identified.

In order for the plan to be successful, there must be an owner who has the responsibility of managing the BCP. The owner must also have sufficient authority to make business decisions for allocating resources to BCP related tasks, evaluating risks, and authorizing spending for BCP. Hands-ion tasks such as documentation may be delegated to others but the owner has the responsibility to make sure the plan is maintained, up-to-date, and tested. In many organizations, the owner is someone (or worse, several people) in senior management whose objective is "Just get it done! And don't bother me with the details" and then they delegate all the responsibilities of managing the plan to other folks who do not have the authority to implement the plan nor approve the associated expense. The result is similar to item #1 where you end up with a great looking set of documentation with no value behind it. Sometimes, the owner is an IT person and then we get into the issues described in item #2 where the plan is heavily focused on DRP rather than looking at the entire business.

4. Inadequate BCP training

It is critical that all members of an organization are familiar with and have been trained on their role(s) within the BCP. Just reading the documents may not be enough. Some roles may require using different systems or accessing information from alternative sources or being familiar with alternate locations. You need to be sure that in the event of a real emergency, there is no confusion about what to do. Also with all the changes going on in businesses all the time, it is very important to train new people on BCP. This is an ongoing process that must be kept on the front burner at all times.

5. Inadequate BCP testing.

The effectiveness of the BCP is dependent on the level and frequency testing. It is not different than fire drills or emergency evacuation drills; you need to be familiar with the process and the best way to do this is to practice a simulated disaster drill. Organizations which do not fully embrace the concept of BCP typically do the minimum of testing to keep costs down and not take people away from doing "real work". The result is that during a real emergency, you may find yourself in a situation which was not anticipated and as a result, precious time now needs to be spent figuring out how to resolve issues rather than executing the solutions. Pay now or pay later; the more testing is done up front, the better prepared you will be.

6. Documentation is out of date.

BCP documents are living documents which take time to put together and time to review for accuracy. This time needs to be invested at regular intervals to ensure the documents still have correct information. If the BCP is write, signed and filed away never to be looked at again until there is a disaster 5 years later, chances are that some if not most of the information is out of date. Organization trees, phone numbers, addresses, suppliers, customers, account IDs, passwords all change all the time. It is critical that as part of BCP, there is an allowance for time to be spent on a periodic review of the BCP to ensure it is current. This can be annual, quarterly; whatever your organization determines to be appropriate but it must done or the value of the plan quickly decreases over time.

7. Inadequate time spent in the planning and evaluating of the BCP strategy.

To create a truly optimized and value-added BCP requires time and effort to analyze and evaluate information from many sources. This is not something one person working by themselves can just zip out to get it checked off the list. Even a large team of expert consultants cannot do this overnight by themselves. They must follow a careful process of interviewing all the key information holders in the organization to ensure all the critical factors are considered before making any recommendations for an effective BCP. These factors include the critical business process, risk factors, impact analyses, recovery time objectives, costs, logistics, and people issues (not necessarily in that order). If any of these factors are skipped over or information is just assumed, then there is a real risk that the plan may not be adequate or it may cost more to implement that is really needed. Nether outcome is acceptable.

8. Inadequate assignments of responsibility.

One to the key success factors to an effective BCP is to assign and distribute responsibilities appropriately. If all critical responsibilities are assigned to one person, then the system will quickly get overloaded and the process will slow down, or worse some tasks will fail. If no one is clear who is responsible, then there will be conflicting and unclear direction or no one will be addressing key issues. There needs to be clear BCP leadership at a top level and then a logical distribution of responsibility at the tactical level. One key item to consider is where there are multiple locations involved; does it make sense to have the BCP structured by function even though the functions cross different locations? Or by site so that there is a local person available to provide direction as it is needed? Also for every assignment of responsibility, there needs to be a primary person and a backup assigned. This ensure coverage of this area in the event the primary person in unable to perform this function.

Having this knowledge provides very valuable insights to enable the creation of a robust yet cost-effective BCP.

How To Get Started Setting Up An Effective BCP

So how do you get started setting up a BCP if you never had one before? And if you do not want to exhaust your budget or profits?

You can do this with or without outside help but you do need a basic outline of the key areas you need to cover. One way to do this is to hire a consultant to help guide you through this exercise. This may run into some big$$$ but it is important and it may be well worth the investment to ensure your continued success in the future. Just one word of advice on this or any other plan for setting up a BCP – please ensure that everyone is on board with this and takes this seriously and understands the importance of the BCP. I have seen too many organizations which just pay a company to come in and set up a BCP so they can say they have one and check this of the to-do list. The problem is that everyone involved within the organization took this as an annoying extra job to do and just gave it a half-hearted attempt to provide any valuable ideas or information. The result is that the organization now owns a $10,000+ set of excellent looking documents which may or may not help in the event of a disaster – but they will never know until a disaster happens because they do not want to bother to spend too much more time testing it nor keeping it up to date. This is really a shame to see good money spent this ay.

The other option, which is better suited to small businesses, is to start with a template and an outline of the key areas which need to be evaluated in creating the BCP.

The template will guide you through all the key phases on BCP creation, such as:

- Organization charts.
- Roles and responsibilities.
- Communication plans
- Business risk analysis
- Business impact analysis
- Plan Activation
- Plan Deactivation
- Testing
- Review
- Approvals

An explanation of an effective plan follow.

Plan Overview and Contact Information

1. Plan Summary

The purpose of the Plan Summary is to identify the name of the company or organization for which this plan applies and to explain the purpose of the plan (what it is designed to accomplish).

The plan summary in the template summarizes all of the key areas that the plan addresses. This is important for presenting this plan to auditors and regulatory authorities. They will want documentation to provide evidence that you have covered all the important bases in your business. Critical elements to consider in implementing a BCP plan include the items listed below, as applicable to your particular business and situation:

- Data back-up and recovery (hard copy and electronic).
- Continuation of all mission critical business operations.
- Operational assessments (impacts to business resulting from interruptions to key business operations).
- Alternate communications between customers and the business.
- Alternate communications between the business management and its employees, contractors and suppliers.
- Alternate physical location of employees.
- Regulatory reporting.

2. Plan Approval
All BCP plans must be approved by at least one individual in a senior company management role. This provides evidence that the business is committed to enacting and enforcing the BCP plan (it is just not a stack of papers that they asked someone to write up and print out to satisfy the auditors).

3. BCP Leadership Roles and Responsibilities
Every BCP plan must have specific roles and responsibilities assigned and agreed to before the plan is approved. In the event of an emergency, each person must already be aware of what their responsibility when the BCP plan is activated or the plan will fail. Once an emergency occurs, it is too late to try to figure this out.

BCP teams are typically broken out by functional area (i.e. Sales, IT, Production) or by site (i.e. Building 1, Building 2, or First Floor, Second Floor, or Springfield office St. Louis Office…). This depends on the size and structure of your organization. If you have a small organization, everyone can be on one team. The goal is to keep the tasks manageable such that the team can be activated and perform their duties quickly and effectively in the event of an emergency. If the team is too big or too spread out, this may lead to time delays due to communication lag time and taking on tasks that are too big. Ultimately it is up to the owner of the BCP to decide how to structure the teams. These are just points to consider.

The key responsibilities to consider in setting up the teams are the leadership assignments and responsibilities, and identifying team membership for each person in the organization. Before an emergency occurs, each person must know who they are to report to and each leader must know who they are responsible for.

The key leadership responsibilities must include:

- Who is responsible for activating and deactivating the BCP plan. In the event of an emergency, there must be one primary person who has this responsibility. Activating the BCP plan has significant financial and operation impacts to the business and only a person authorized to make this call should activate the plan. This person must be in a position of authority to have the business information they need in order to be able to make an informed decision to activate the plan. They must also be in a position to quickly receive notification of any events which could trigger the BCP (i.e. Fire, Massive power outage, earthquake, chemical spill, etc.).
- Who is responsible to provide leadership and decision making authority to BCP team upon activation of the BCP plan, until plan is deactivated. This can be the same person who activates the plan and leads the team. This is up to you to decide, but you must spell this out or else there will be mass confusion when an emergency strikes. People need to know clearly

who to take direction from. There needs to be one identified authority to do this.

- Who will manage external communications between the company and outside parties and internal communications between the BCP team and company personnel. This is the person who receives and sends information to and from others inside and outside the company. This includes sending/receiving information to/from local emergency management personnel/police/fire department/etc. and for overseeing communications of BCP/emergency information to/from company personnel. In the event of an emergency, there musty be a clear line of communication and everyone needs to know where to go to accurate and clear information. Without planning this critical piece of the BCP plan, the resulting frenzy of informal communications will cause massive confusion and panic. This needs to be avoided or else the plan will fail.
- Who is responsible for the BCP plan itself; the BCP plan administrator. This is the person who owns and manages the BCP plan to ensure it is current and accessible in the event of an emergency. This is a key role which needs to be taken seriously as a primary responsibility to get this plan in place and keep it up to date. Ensure review and testing is performed and documented on a regular basis. This means the plan is checked to make sure the information is still current and accurate. Also make sure that the plan is tested out to make sure it will work and to find out where revisions or additions need to be made to fill any gaps uncovered during testing. This is not just something do when time permits, or else it will not get done or it will get out of date and then when an emergency strikes, you will be out of luck trying to scramble to find out what information is current and what is not. Time is money and the time you pout in up front to get this plan in shape will pay off later on when you need it to be correct the first time you activate it.
- Who is in charge of each sub team (Functional Lead). These are the people who are responsible for leading each of the BCP teams as described above.
- VERY IMPORTANT: For each of these roles described above there must be a primary and a backup person identified. In the event that the primary person is not available or unable to perform their role, the backup person must take that role. The backup person must be made aware that they have been designated as the backup person for their specific role before the emergency happens. They must have access to the same information as the primary person, has and they must take part in all BCP training and testing exercises so that if needed in an emergency they can step into the primary role without any lag time.

4. Internal Contact Information

In the event of an emergency or disaster, it is critical to have contact information for all personnel to be able to quickly share important information as it becomes available and to quickly account for all personnel. The most useful format for this information, in the event of an emergency, is to set it up as one or multiple contact trees (depending on how big your organization is).

The contact tree is a structured hierarchical format with multiple levels which look like a pyramid if you draw them in a diagram.

The critical pieces are:

- Each person is assigned a specific box in the contact tree.
- There are defined levels to ensure clear direction of communication.
- The tree is designed to map a logical grouping of individuals (i.e. by function, by department, by site).
- The tree is designed to distribute the tasks of making contact (i.e. no one person does all the work). Ideally each person should have 1-5 contacts to make.
- Each box is clearly associated with levels above and below.
- Each person knows who will contact them, and in turn who they will contact.
- Multiple means of contact are identified for each person (daytime and off-hour phone, cell phone, email, etc.) so that if one means of contact does not work another can be attempted.
- Special skills are identified which could be vitally important in an emergency situation (i.e. who is a doctor, paramedic, fire fighter, etc. who can be asked to help tend to any urgent situations).
- There is a confirmation that loops back up the tree to ensure that all the people in the tree are accounted for, or to alert the top level that people cannot be accounted for. It is not enough just to call down the tree and presume that everyone has gotten the message; you need to know for sure that everyone has been reached.

This is one of the most important pieces of the BCP. Without basic communications, the rest of the plan will fail. Be sure this section can easily be printed out and provided to all personnel in the tree. In the event of an emergency, each person needs to access this plan quickly and must not be dependent on any electronic or online systems which may not be functional. A .pdf copy is not sufficient if the power is out or you cannot get to your computer.

This is how the contact tree works -

- In the event of an emergency, the BCP leadership contacts the person at the top of the tree (Level 1).
- The Level 1 person then contacts the assigned people in the tree one level down (Level 2).
- The Level 2 people then contact the people assigned to them one level down (Level 3).
- Once the Level 2 people have contacted (or attempted to contact) all the Level 3 people assigned to them, they then contact the Level 1 person

above them to report the status (All People Accounted For or People Not Accounted For) and any question or information to be shared.

This provides confirmation that the contact tree has reached all the affected individuals and that everyone is accounted for. The Level 1 person at the top of the tree is then responsible for reporting back to the BCP Leadership the status of people in their contact tree and conveying any information they need to share.

The contact tree can be set up various ways depending on what makes sense to your organization such as supervisor with associated personnel below, or it make more sense to have a level 1 site person with assigned people at the same site below (for organizations where reporting structures span multiple locations).

The contact tree can be expanded (more levels added) for larger groups. Multiple contact trees can be set up for multiple groups, functions, or locations.

The table provided in the template is set up in a contact tree structure to identify each person within a contact tree and to provide instructions on what to do in the event the contact tree is activated. Information about each person should include how to reach them through multiple means (phone, email, text, etc) and at different times (work hours, after work hours, etc). It is also very useful to note any special skills or training each person has (i.e. paramedic, firefighter, doctor, nurse, etc.) in the event their services are needed in an emergency.

5. External Contact Information (Customers)
This section documents contact information for your customers. In the event of an emergency, you must be able to continue to communicate with your customers in order for the business to continue. To be prepared for this, you need to have a list of your customers and key information. This should include at a minimum their name and contact information and information about any orders/contracts they have with your business. This list can be documented in the table included in the template or it can be a report from one of your systems. Depending on how big your business is, you may choose to have only the very top customers' information documented in portable format (i.e. printed out so you can take it with you) and have the rest of the customers' information saved in a computerized system (which has a backup/disaster recovery plan in place to ensure you can access a copy of this data in the event that the primary system is not functional). The bottom line here is that in the event you need to contact a customer or client in an emergency situation, you can still do it.

6. External Contact Information (Suppliers and Business Service Providers)
This section documents information you will need to keep in contact with your service providers and suppliers. These include suppliers of services or materials needed by your business, legal services, business services, etc. These contacts will be especially important in the event of an emergency (i.e. contacting the insurance company in the event of disaster or notifying your suppliers of alternate

work locations) so it is important to have the information easily accessible. It is best to have this in one list, which includes the names and contact information as well as account or policy numbers you will need to identify yourself. It is not advisable to rely on multiple sources for this information of old documents (i.e. bills, brochures, quotes) as these will take time to find or may not be accessible at all during an emergency. One option would be to scan these documents and create a single .pdf which you can copy (keep a backup in an alternate location) and print out to keep with you. Another would be if you have this information in a database or computer system, run a report which summarizes the information you need and print this out to keep with you.

7. Utilities and Facilities Services

This section is very similar to section 6 but focuses on utility companies and facilities services such as leasing companies. It is very important to have this information easily at hand in the event of an emergency where you might need to contact the utility company to shut off service or report an outage etc.

8. Financial Services Contact Information

This section documents the information you will need to communicate with financial services (i.e. Bank, payroll and benefits services, etc.) and access your accounts. In the event of an emergency or disruption to business requiring relocation, it is imperative to maintain contact with financial institutions to enable continuation of payments to employees, suppliers and service providers, as well as processing payments from customers.

The list must contain the names and contact information for all of your financial services so that you have this in one place easily accessible in the event of an emergency. Sensitive information such as account numbers, log on IDs, PINs and passwords are important to be able to access your accounts but you probably do not want these printed out on a list that some can copy and use to get into your accounts. So the best way to manage this is to designate key individuals in your organization who have these pieces of information and are responsible for managing them and keeping them confidential. For purposes of the BCP plan you then need to identify the individuals (make sure you have a primary and a backup person identified) who is responsible for managing the account and password for each financial service. In the event of an emergency, this person would be contacted and asked to access the account and perform whatever management needs them to perform to keep the business operating. This is another reason why the contact tree is so vitally important because you want to be sure you can contact these people.

9. Regulatory Agency Contact Information

This sections documents contact information for regulatory agencies and authorities such as IRS, Tax departments, local authorities and other regulatory bodies depending on your business (i.e. .FDA, EPA, DEP, Customs, Licensing agencies, etc. as applicable). It is critical to maintain communications with regulatory agencies during emergencies to enable reporting of business status affecting required regulatory filings or reporting time sensitive business critical information to regulatory agencies. You don't want to add to your troubles by getting behind on renewing your license to manufacture your product or paying your taxes because you cannot locate the information you need because it is on a computer which you lost in a flood.

10. Critical Records and Systems

This section identifies your business critical records and systems. This includes paper files and compute system which contain business critical data (such as orders, financial transactions, product specifications, product manuals, software source code, etc.). For all of these records and systems you need to have a back up system and/or location identified and documented in a list that is easily accessible in the event of an emergency. Backup locations and systems include paper copies, microfilm or scanned copies of paper documents stored at an alternate location, and system data and software backed up on DVD's or tapes etc. and stored at an alternate location.

It is not necessary to document system IDs and passwords here (unless you want to) but at a minimum you need to identify who has the necessary system IDs and passwords, pin, keys, combinations, etc. to be able to access the records/systems. Be sure to identify a primary and backup person here.

Ideally backup copies of systems and data should be copied onto multiple media. Most modern libraries that backup irreplaceable data use multiple media such as optical media (CDs, DVDs) plus magnetic media (Tapes, Hard Drives) and solid state media (Memory Stick, Jump/Flash/Thumb Drive). With the emergence of cloud computing, the new trend is towards online storage which is accessible anywhere on the Internet. Services of this type are provided by companies like Amazon or Online Mailbox. The important piece to consider is that you need to be able to access important information and you should not rely too heavily on just one media because there are multiple causes of failure for each type (i.e. Tapes can be demagnetized, CDs can warp if exposed to heat, Jump drive will fail if they get wet, etc.).

For all of the storage media options and for copies of paper document, at least one copy of these needs to be kept in an alternate location away from your primary work location just in case you cannot access your primary work location. Having you critical files in a fireproof safe which is in your office which you cannot get to because of an evaluation due to a massive chemical spill is not going to help you. You need a copy that is either with you at all times or in a location away from the primary location. The general rule is that the backup location should be approx 20-50 miles away from the primary location (close enough so that you can get to it easily but far enough away so that it is not affected by the disaster). The back up location should ideally be protected from theft, fire and water damage and should ideally be climate controlled (temperature and humidity are controlled to minimize damage to your records and media). There are companies such as Iron Mountain who provide this type of service.

11. Backup Locations

This section documents the backup work locations identified which are to be used in the event of an emergency which prevents work from happening at the primary location. Some tasks can be performed remotely such as in a home office (i.e. Sales functions or technical writing) but others will require backup locations (i.e. manufacturing, laboratory work). Identification of backup locations and agreement on the terms of using the backup locations need to be defined and documented before the BCP plan is finalized.

These considerations include:

- Space requirements vs. availability
- Qualification/certifications required to perform the work at the backup location
- POs /payments/contracts of financial agreements required
- Time needed between notification and occupying the backup location
- Using internal personnel at backup location or outsourcing the operation entirely

The purpose of this section is to ensure continuation of critical business functions during activation of the BCP plan. During an emergency you want to know who to contact to get business operations moving quickly. You cannot afford to be researching options for resuming business operations. Backup/contingency locations must be identified for all essential operations which cannot be performed remotely (at home). Backup sites should be far enough away from the primary work site to reduce the chances that the disaster will affect access to the backup site, but not so far as to prevent timely access by personnel. A general rule is for the backup site to be 20-50 miles from the primary site. Backup site information in the BCP plan must include directions instructing personnel how to get to the location.

Directions and Maps to all backup locations need to be included in the Appendix Section of the BCP to be able to easily provide copies to all personnel who need to report to the backup locations.

12. Backup Service Provider and Supplier Information

This section documents a list of your emergency/backup suppliers. These are the people or companies you would go to in the event that you cannot use your normal or primary suppliers either because you have moved work locations during an emergency or because there is an emergency at the primary supplier which makes them unable to help you.

In the event of an emergency or disruption to primary business services and suppliers, a list of back up services and suppliers must be maintained for quick reference to enable restoration of essential business functions in a timely manner. This list should include backup suppliers for essential functions necessary for the business to operate, such as computer services, shipping, order processing, etc.

13. Emergency Services Contact Information

This section is a basic list of emergency services such as police, fire, etc. In any emergency situation, it is essential to know who to contact. This information must be on hand for easy availability. List all critical emergency contact information in the table below so you have it all in once place for easy access.

<u>Business Risk Assessment and Impact Evaluation</u>

This section of the BCP plan documents the risk associated with the disruption of any key business operation or process. Risk is defined as the combination of how likely a key operation will be disrupted, how much time before the business experiences the negative impact of losing the operation, and how much this disruption will hurt the business performance. (Ex. The manufacturing plant can be shutdown for several months in the event that the river running through the property overflows and floods the building. Some flooding has occurred once every 3 years with major flooding occurring on average every 25 years. This will result in loss of sales within 2 days).

The information documented in this section is used to justify the prioritization of the BCP plan activities and justify the money allocated to ensuring the continued operations during an emergency or disaster. (Ex. A backup manufacturing site is qualified and PO-##### is issued to Company Z to enable start up of production within 1 day in the event of an emergency at the primary manufacturing facility. A new manufacturing site has been identified and a project initiated to relocate primary manufacturing to this site by 15-JUN-20##).

The more these figure can be quantified, the better will be the BCP decision-making.
To assist in this evaluation, use the scale indicated in the chart below.

Once this section is complete, this information will provide a roadmap for prioritizing the BCP tasks.
The result will be a list of key business operations/processes with risk levels assigned. The operations/processes with the lowest risk numbers will have the highest risk level. The high risk items are the ones which should receive the most attention when designing and implementing the BCP plan. These are the items which will hurt the business the most if they are not adequately addressed.

14A. Business Process Risk Evaluation

This section documents the risk associated with each business operation or process. In order to create a BCP plan that is effective and manageable it is important to understand the relative risk associated with an interruption to each critical business operation so that the BCP plan can address restoring these operations in a prioritized ordered approach. It is impossible to go from a disaster which interrupts your business to bringing everything back online all at once. There is so much to manage and many limitations during emergencies that you need to be very clear before the emergency happens what your priorities are. If you lost your primary work site, what operation/function would you need to bring back first? Second? Third? Etc.

The critical elements of completing this section are:

- Identifying the key business operations and processes. These are operations or process which must be working in order for the business to continue. Examples are customer service, processing orders, shipping product, etc.
- Identifying the likelihood of disruption (chances that something unexpected could happen which disrupts the key business operation or process). SCALE: 1 – 10 (1 = Very High Likelihood of Unplanned Disruption, 10 = Very Low Likelihood of Unplanned Disruption)
- Identifying how significant the effect is of losing this operation or process. (i.e. people at risk, profits lost, customers lost) SCALE: 1 – 10 (1 = Very High Significance To The Business, 10 = Very Low Significance To The Business). Identifying how long this operation can be disrupted before it begins to negatively impact the business. Number of Days without this operation before business would experience any negative impact to operations and/or sales.

14B. Business Process Disruption Impact Evaluation
In this section, the impact to the business of disruptions must be evaluated for each critical business operation identified in Section 14A. The object here is to define how much of an impact will disruption of each critical business process have on the business.
This must be defined in terms of:

- Available work-arounds to temporarily replace the process (ex. paper records can be processed manually in the event that a key computer system is unavailable, process can be outsourced to Company X).
    - A workaround is a manual process or other way to accomplish what needs to be done in the event that the primary process is unavailable. For example, you can take orders on paper forms if the computer system is down or you can manually produce a product if the automated production line is down, etc.
    - If there is a workaround, there will be less impact to the business as the BCP plan should specify that the workaround is activated in the event that the operation/process is disrupted. The BCP plan should include instructions on how to perform the workaround.
    - If there is not a workaround, the impact to the business will be greater because when the primary operation/process is interrupted

your business will feel the effects (i.e. sales lost, no more orders taken, etc.)

- How long can the business continue to operate until the process must be restored? (ex. computer system must be back online within 3 months or costs and loss of efficiency will reach unsustainable levels).
    - This is not the same as the question in section 14A (how long before the business begins to feel the effect of a disruption to the operation). This is how long before you decide the process must be restored. This is different because due to priorities and economic/resource limitations, you need to make the proactive decision to wait a certain period of time before taking action to restore this operation due to the cost and resource considerations this will have vs. the risk of waiting. The information documented in Section 14A (how long until the business begins to feel the negative effect to the operation) is a key piece of information needed to make an informed decision on how long to wait until restoring the operation. Other operations may take priority.
    - This needs to be decided and documented before an emergency happens.

- What resources are needed for the work-around and for restoration of the process?
    - In order to effectively plan for restoring a critical business operation, you need to know what resources (people, time, materials, supplies, money) are needed.
    - This information needs to be documented in advance of an emergency so that when you need to activate the BCP plan you know right away what you need. Even better, you can put together some of these supplies and materials ahead of time and have it ready just in case.

List the Key Business Processes in order of BCP Priority (lowest number first) using the information identified in the table in question 14A. The objective here is to identify which processes require the most immediate attention in the event of a disruption and what is required to address the disruption.

Pre-Emergency Planning

15. Risk Mitigation

The best way to prevent getting caught in a disaster is to take proactive action and see what you can do to prevent a potential disaster or at least minimize the impact. For example if you identified that storing flammable materials, such as spray paint and paint thinner, near your hot water heater (which is heated by gas and has an open flame) could result in starting a fire, your mitigation action would be to move the flammable materials to a safe location to reduce the risk of a fire starting.

This same logical approach needs to be taken with your business. If you find that you have data that is not backed up or if you have computer systems which are vulnerable to viruses or unauthorized access, you need to either determine that the risk is acceptable or take action now to reduce your identified risk as best you can. Not every risk must be mitigated. Some are not possible to mitigate so they need to be planned for in your BCP, some are too costly to mitigate, and some may not carry a high enough risk for you to spend resources on mitigating. Only you can determine this, but you need to go through this exercise so you can make informed decisions and plan accordingly.

16. Emergency Materials and Supplies

To prepare in advance for an emergency, it is vital to have emergency supplies on hand and easily accessible to the people who need them. This should include critical emergency materials and supplies that need to be available in the event of an emergency or unplanned business interruption. This includes small disposable supplies (i.e. pens, paper, mailing supplies, etc.), office equipment (computers, printers, phones, etc.), production equipment (i.e. machinery needed to service customers, refrigerator, freezer, generator, batteries, tools, etc.) as well as raw materials and production supplies (i.e. paper, paint, lawn fertilizer, baking ingredients, gasoline, propane, etc.).

In the previous section, you identified what supplies are needed to restore each business operation. Now that you know what you need, you must decide how much of this you will acquire and keep on hand before an emergency happens. This is determined based on the cost of keeping in emergency supplies on hand versus risk of not having them. Only you can determine this.

Some supplies can be easily obtained if needed in a hurry but others are more difficult. During a flood, I know from experience it is impossible to find sump pumps or generators. Even the large suppliers like Home Depot and Lowes ran out of sump pumps and generators and did not expect more for another week, and they had a waiting list of people wanting them. Meanwhile the water level was rising in my newly finished basement. Consider this when completing this section.

BCP Plan Activation and Implementation

In this section, identify the triggers for activating the BCP plan (what events must occur in order to activate the BCP plan), who is involved in activating the BCP plan, and what are the steps to activate and implement the plan. All key business processes identified in Section 14A should be listed in Section 16 in order of BCP priority

Start times should be specified based on impact evaluation information captured in section 14B (ex. if production line downtime causes loss of customer orders within 2 days, then trigger is that production line is down for more than 1 day and target start time for production operation BCP should be specified to have a backup process in place within 2 days).

17. Plan Activation
This is the section which needs to contain the specific trigger for activating the BCP plan.
It must clearly identify:

- What needs to happen in order for the BCP plan to be activated?
- Who is responsible for activating the BCP plan (the primary and backup person)?
- Which BCP plan (the instructions for addressing the disaster) are to be followed?
- Which contact tree needs to be activated to notify the proper individuals of BCP activation?

It is disruptive and can be expensive to activate a BCP plan, so you need to be clear up front what situations warrant the activation of the BCP. Activating a BCP is not appropriate for every interruption. For example, if there are downed wires that knock out power for 6 hours to your primary manufacturing site, this may or may not be enough to activate the BCP. You need to decide this based on the evaluation performed following the template. This depends on how long you can be without the operation versus how long it would take to get the backup operation going and how much it would cost. You may decide you can afford to wait for up to 2 days without power before activating the BCP. But you need to do this evaluation and document the trigger up front before the emergency happens. If you know you can wait 2 days, then the appropriate action is to start counting the time from when the power is interrupted and at 2 days you need to activate the BCP plan if the power is not coming back on.

Having a plan is not enough; you need to be sure when to use it in order for it to be effective otherwise you could be wasting time, effort, and money by needlessly panicking for no reason.

18. Plan Implementation
This section provides the instruction for carrying out the BCP plan. It needs to be very clear and easy to follow. It is highly recommended that one plan is written for each critical business [process identified in the template. The plans should be written clearly in stepwise format (i.e. Do Step 1, then Step 2, then Step 3). For each step there needs to be a person assigned to do this step, so one assumes someone else is taking care of something (and then no one does). It is also good practice to have a signoff for each step so you have written confirmation that a step has been completed. This is especially helpful in reviewing BCP implementations and tests to see what worked well and what needs

improvement. This is also very helpful for BCP plans which involve handoffs (i.e. person A completes Step 1 and then Person B completes Step 2). The tables in the template are set up to document the tasks, who is responsible and a section for verification that the task was completed. As an alternative to these tables, the implementation plan can be a separate document attached to the BCP Plan as an Appendix.

BCP Plan Deactivation and Restoration of Normal Operations

This section defines what happens once the emergency situation has passed or is no longer an impact to the business. The objective here is to return to normal business operations. First you will need to define what must happen in order for you to begin the process of returning to normal business operations or to a longer term solution (the BCP deactivation trigger), and then you need to define what the instructions/steps are to deactivate the BCP and return to normal operations (deactivation implementation/return to normal operations).

This section is set up just like the BCP Activation and Implementation section above but it is to provide instructions on the opposite process. Your objective is to be able to get back to a normal mode of business and you need to know when and how to deactivate BCPs as soon as it is appropriate.

19. BCP Plan Deactivation

In this section, document the BCP De-Activation Trigger for each Key Business Process; what would need to happen in order to stop following the BCP plan and go back to conducting business like you did before the emergency (Ex. Utilities at primary work location have been restored). Be sure to document who is responsible for deactivating the BCP and initiating the restoration plan. Also be sure to document the affected Contact Tree (i.e. who needs to be informed of the deactivation of the BCP and initiation of the restoration).

20. Return To Normal Operations

In this section, define the tasks needed to move from operating under the BCP plan to restoring normal operations. For each task, document who is assigned the responsibility for completing the task. The tables in the template are set up to document the tasks, who is responsible and a section for verification that the task was completed. As an alternative to the tables, the restoration plan can be a separate document attached to the BCP Plan as an Appendix.

BCP Plan Testing

To ensure that a BCP plan will be effective in the event of an emergency, it is important to test it out. Just like a fire drill tests out how good your plan is for evacuating the area in an emergency and how well people are trained and prepared to respond to an evacuation, you need to know how well your BCP plan is set up and how well people can follow it. Testing will identify any areas that need attention or revising.

The degree to which the testing is conducted is based on what you feel is appropriate. Obviously you will not set your office on fire to test the BCP, but many organizations do conduct emergency backup testing of computer systems (disaster recovery testing to demonstrate that the system can be brought up at the backup site and it is still accessible and still works and provides the data needed to operate the business). Other tests include simulations to varying degrees. Some examples are having everyone work remotely for one day to verify how well the networks and systems operate and note any problems. Or do some laboratory testing or manufacturing at the back up site to verify how well and how quickly and how productive the backup plan works.

The objectives of this section are:

- To demonstrate that the plan does work.
- To provide people an opportunity to be trained in the BCP so they ready for a real emergency.
- To identify any problems, issues or areas of improvement that need to be addressed in the plan.

As a result of testing you may find out that the back server needs more disk space, or the data backups need to happen more frequently or people working remotely may not be able to access necessary systems because of firewall issues, or your backup manufacturing site does not have a loading dock so all you materials needs to be carted using fork trucks. The only way to realize these potential problems is through testing.

Make you sure you at least test the contact tree on a regular basis. You should be able to contact all the people in the tree and get confirmation back. Note how long this takes and any difficulties you encounter. These need to be addressed as one of the most basic and fundamental failures in a BCP plan is not being able to get in touch with everyone involved. If people have important roles that are critical, you might need to get them a better cell phone or faster internet service so you can communicate with them in an emergency. During the Y2K planning, key individuals in the company I worked for were given satellite phones in the event that the cell phone network went down. You need to decide what is appropriate.

Testing should be performed on a regular basis which you must determine based on what you think is appropriate. Some organizations perform yearly testing, or partial testing one year and full testing the next. When significant changes happen (you use a new back up site for your servers or take on a new manufacturing process) you would probably want to test the BCP to ensure these work as you intend.

Ideally, the test should be documented in a report which is attached to the BCP plan to document what was done, what the results were and any follow up actions that need to be taken to ensure effectivity of the BCP plan.

BCP Plan Maintenance
It is important to keep a BCP plan current and up to date in order for it to be useful when you really need it. This means that the plan must be reviewed and information updated on a regular or as-needed basis. This can either be done every year or every 6 months as it makes sense, or it can be every time there is a change that impacts on the information contained in the BCP plan.

The review should look at all the critical processes and make sure none are missing or have changed significantly such that the BCP plan no longer is adequate or accurate. You also need to carefully review contact trees to make sure the right people are in the right places in the tree. And you need to ask people to confirm that their contact information is still correct.

When the plan is reviewed, indicate the date the plan was reviewed and what changes, if any, were made. If the plan is reviewed on a regular time basis, then it is possible that the review will find that all of the information is current and correct and thus no changes are necessary. It is still important to indicate the date and completion of this review to specify that the plan is still current.

Appendices / Attachments
Appendices or attachments are supporting documents that are attached to the BCP plan document. It is important to have these documents on hand in the event of an emergency. By attaching copies of them to the BCP plan, you will have these all in once place in case you need them and cannot access your regular work location.

Examples:

- Map to alternate locations
- List of suppliers
- Contact list for temporary resources
- Customer list
- Instructions on how to set up a remote office network
- List of IDs and passwords for online systems
- BCP Plan test reports

Emergency Planning for Personnel and Families

In case of emergency or disaster, it is important that you and your family has a plan so you know what to do and do not panic. Before an emergency happens, please take the time to record the vital information listed below. Make several copies of this plan and save them in safe places so you can access them at all times whenever you need them. Keep one with you at all times, and keep another in your emergency supply kit. **Make sure you review this plan on a regular basis to ensure the information is plan up-to-date.**

Out-of-Town Contact Name:	Telephone Number: Cell Number: Email:
Local Meeting Place:	Address: Telephone Number:
Alternate Meeting Place:	Address: Telephone Number:
Evacuation Location:	Address: Telephone Number:

Information about Family Members.

Name:	Location during the day:
Date of Birth:	Important Medical Information:
Name:	Location During the Day:
Date of Birth:	Important Medical Information:
Name:	Location During the Day:
Date of Birth:	Important Medical Information:
Name:	Location During the Day:
Date of Birth:	Important Medical Information:
Name:	Location During the Day:
Date of Birth:	Important Medical Information:
Name:	Location During the Day:
Date of Birth:	Important Medical Information:

Locations during the Day:

Work Name:	School Name:
Address:	Address:
Phone Number:	Phone Number:

Work Name:	School Name:
Address:	Address:
Phone Number:	Phone Number:

Work Name:	School Name:
Address:	Address:
Phone Number:	Phone Number:

Other location:	Other location:
Address:	Address:
Phone Number:	Phone Number:

Important Information	Name	Telephone Number	Policy Number
Doctor:			
Doctor:			
Doctor:			
Doctor:			
Dentist:			
Pharmacist:			
Medical Insurance:			
Homeowners/Rental Insurance:			
Automobile Insurance:			
Child Care Provider:			
Child Care Provider:			
Veterinarian / Kennel:			

Dial 911 for Emergencies

Inventory of Emergency Supplies

Item	Location	Expiration Date	√ *
Water, one gallon of water per person per day, for drinking and sanitation			
Prescription medications and glasses			
Food, at least a three-day supply of non-perishable food per person (canned food, dehydrated food, energy bars).			
Infant formula and diapers, if you have an infant			
Pet food and extra water for your pet			
Cash or traveler's checks and change			
Emergency reference material such as a first aid book			
Cash or traveler's checks and change (credit cards may not work if there are power outages or interruptions in phone or internet lines)			
Feminine supplies and personal hygiene items			
Books, games, puzzles or other activities for children			
Local Maps			
Battery-powered radio and a Weather Radio			
Extra batteries			
Fuel for cooking and heating			
Gasoline (stored in approval safety containers, not inside the house) for transportation or operation of generator or emergency equipment			
Flashlights			
First Aid kit			
Whistle to signal for help			
Moist towelettes, garbage bags and plastic ties for personal sanitation			
Dust mask or cotton t-shirt, to help filter the air			
Plastic sheeting and duct tape to shelter-in-place			
Wrench or pliers to turn off utilities			
Can opener for food (if kit contains canned food)			
Fire Extinguisher			
Paper & Pen/Pencil			

Matches in waterproof container			
Household chlorine bleach. (You can use bleach as a disinfectant; diluted nine parts water to one part bleach, or in an emergency you can also use it to treat water. Use 16 drops of regular household liquid bleach per gallon of water. Do not use scented, color safe or bleaches with added cleaners.)			
Important Family Documents such as copies of insurance policies, identification and bank account records in a waterproof, portable container			

*Ensure all items are available in sufficient quantity and not past expiration date.

Printed in Great Britain
by Amazon

30763547R00073